ISABELL

The Bird Man

The Extraordinary Story of

John Gould

First published in Great Britain in 1991 by
Barrie & Jenkins Ltd
20 Vauxhall Bridge Road, London SW1V 2SA

This edition published by Ebury Press in 2004

British Library Cataloguing in Publication Data

Tree, Isabella
The Bird Man

ISBN 0091895790

Typeset in Imprint by SX Composing Ltd, Rayleigh, Essex
Printed and bound by Bookmarque Ltd, Croydon, Surrey

For Nancy Mam

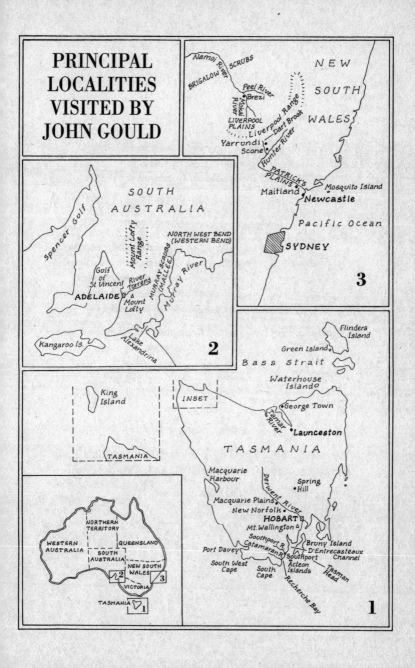

PRINCIPAL
LOCALITIES
VISITED BY
JOHN GOULD

Map 3 — New South Wales

Namoi River
BRIGALOW SCRUBS
NEW SOUTH WALES
Peel River
Brezi
Mokai River
LIVERPOOL PLAINS
Liverpool Range
Dart Brook
Yarrundi
Scone
Hunter River
PATRICK'S PLAINS
Maitland
Mosquito Island
Newcastle
Pacific Ocean
SYDNEY

Map 2 — South Australia

SOUTH AUSTRALIA
Spencer Gulf
Mount Lofty Range
NORTH WEST BEND (WESTERN BEND)
Gulf of St Vincent
River Torrens
MURRAY SCRUBS (MALLEE)
Murray River
ADELAIDE
Mount Lofty
Kangaroo Is.
Lake Alexandrina

Map 1 — Tasmania

Flinders Island
Green Island
Bass Strait
Waterhouse Island
King Island
INSET
TASMANIA
Tamar River
George Town
Launceston
TASMANIA
Macquarie Harbour
Spring Hill
Derwent River
Macquarie Plains
New Norfolk
HOBART
Mt. Wellington
Southport R.
Catamaran R.
Port Davey
Bruny Island
D'Entrecasteaux Channel
Southport
South West Cape
South Cape
Acteon Islands
Tasman Head
Recherche Bay

Inset map — Australia

WESTERN AUSTRALIA
NORTHERN TERRITORY
QUEENSLAND
SOUTH AUSTRALIA
NEW SOUTH WALES
VICTORIA
TASMANIA

Contents

ACKNOWLEDGMENTS

My deepest thanks are due to Dr Gordon Sauer of Kansas City, author of a bibliography and chronology of the life of John Gould, who put his entire collection of Gould material at my disposal. It was through his extraordinary generosity that I found myself with copies of hundreds of letters, articles and books – virtually all of the extant Gould material – miraculously at my fingertips. I owe him my thanks for sharing with me the fruits of nearly 50 years' research, and for letting me flog his photocopier within an inch of its life.

I must also thank Sandy Mason and James Helyar of the Spencer Library at Kansas University who showed me selections of original Gould drawings from the great Ellis Collection.

I am grateful to members of Gould's family – Maureen Lambourne, the late Grace Edelsten, and her son David Edelsten – for all their hospitality and help, particularly for permission to use letters, journals and photographs in their possession; and, most especially, the late Dr Geoffrey Edelsten who shared his own notes and collection of Gould material with me.

My thanks, too, to the late Lord Derby for showing me the wonderful collection of original drawings and books in his library at Knowsley; to my sister, Esther, for helping me with research; to Ann Datta of the Zoology Library at the Natural History Museum, for coping with countless photocopy requests and for retrieving letters from the Museum's vast, uncatalogued Gould collection; and to other members of staff at the Natural History Museum, the Library at Knowsley, and the Zoological Society, for heaving about all those hefty volumes from shelf to table.

Finally, I would like to acknowledge the following for granting permission to quote from letters in their collections: The Trustees of the National Museum of Scotland; The Trustees of the British Museum (Natural History); The President and Council of the Royal College of Surgeons of England; the American History of Natural History; the Kenneth Spencer Research Library at the University of Kansas; the Scott Polar Research Institute; the Royal Society of Tasmania; the Royal Geographical Society; Cambridge University Library; the Balfour & Newton Libraries, Cambridge University; the Zoological Society of London; the State Library of Victoria, Melbourne; the Library of the Academy of Natural Sciences of Philadelphia; The Duke of Westminster; the Blacker-Wood Library, McGill University, Montreal; the Mitchell Library, Sydney; the National Library of Australia, Canberra; the New York Botanical Garden Library; the Musée Nationale d'Histoire Naturelle, Paris; the Houghton Library, Harvard University; the Rijksmuseum van Natuurlijke Historie, Leiden.

I would also like to thank Hannah MacDonald, Julian Shuckburgh and my agent, Gillon Aitken.

LIST OF ILLUSTRATIONS

INTRODUCTION

King George IV's Giraffe

JOHN GOULD, author and publisher of some of the most magnificent illustrative work on birds ever to have been created, staked his first claim to fame by stuffing a giraffe. It was not something Gould was inclined to advertise at the time. With his fledgling reputation as a serious ornithologist to think of, he was keen to avoid being branded a mere 'animal stuffer'. And yet preserving King George's favourite pet brought with it unprecedented publicity and gave Gould the recognition he needed to launch himself into the scientific world. For this was no ordinary giraffe: it was a symbol of Britain's scientific awakening; it was a glove cast down before the great scientific establishments of Europe; it was a national monument. When Gould set about his messy, noxious task skinning his unwieldy subject, disembowelling it, extracting its bones, removing its brains, painstakingly scraping the fat from its enormous skin, turning its ears inside out to extricate cartilage, building a substitute skeleton from iron rods and wood, fleshing it out with straw and wads of hemp, stretching the skin over its new body (taking care to keep the mane on that fantastic neck as straight as the seam down a stocking), and finally sewing it together again, he was preparing the mould for an ingenious and extraordinary career. And for this unique opportunity he could thank Mehemet Ali, pasha of Egypt.

Four years earlier, in 1826, the pasha had conceived the brilliant if unconventional idea of sending three live giraffes to France, Britain, and Austria as tokens of friendship and goodwill. It was a gesture as diplomatic as it was imaginative. It was, after all, only 25 years since French troops were hacked to death as they evacuated Egypt following Napoleon's abortive mission in 1798; and it was less than 20 years since the disastrous British expedition into Egypt, when the pasha, a former British ally, had constructed a triumphal alleyway in Cairo with the severed heads of 450 British soldiers staked on the ends of poles. With the Treaty of London declared in July 1826 between England, France, and Russia (Austria had opposed it), as a gesture of protection for Greece against the combined threat of Turkey and Egypt, Mehemet Ali took the opportunity to reassure the European nations of his goodwill and his interest in continuing commercial and diplomatic relations.

The duplicitous pasha, however, was treading a very thin line. His olive branch would have to be dignified, flattering, and, above all, irresistible if it was to be well received by the watchful heads of state. Europe's escalating addiction to zoological novelties from far-flung parts of the world provided him with the perfect solution.

Live animals or birds had long been given as exotic gifts between countries, but never before had traffic in wild animals been so prolific and competitive as during the decades following the turn of the nineteenth century. The dawning of the Age of Science had equipped Europe with an insatiable appetite for novelties from the natural world. While there were a few travelling circuses exhibiting their shabby curiosities to the general public, there were almost no centres in Europe where wild animals were on permanent display.

In Paris the Jardin des Plantes, earlier the Jardin du Roi, held perhaps the most substantial collection. It was partly in an effort to 'keep up with the Joneses' and partly as a result of his own extravagant taste for the bizarre and exotic that George IV of England had been busy furnishing his own 'jardin' at Windsor Castle with curiosities on a scale and rarity similar to the Parisian model. By 1826 the Royal Menagerie at Sandpit Gate boasted a collection of gnus, monkeys, roebucks, kangaroos, musk deer and elks, mandarin horses and brahmin bulls, a zebra, a leopard, a llama and a tortoise.

Elsewhere the Duke of Devonshire kept 'a particularly sagacious female elephant' at Chiswick House; Edward Cross kept lions, a clouded tiger, a leopard and a mandrill called Happy Jerry in his private menagerie known as the 'Exeter 'Change' off the Strand; and Lord Stanley (later the 13th Earl of Derby) was amassing his gigantic living collection at Knowsley near Liverpool. But nowhere, in collections private or public, had there appeared a giraffe.

For some reason, the 'camelopard', as it was first described, had proved particularly elusive. Myths of monster proportions veiled the true identity of this gentle creature. Firsthand accounts and tangible proof in the form of giraffe skins had been surprisingly difficult to come by; and as late as 1810 an albino camel painted with spots was successfully paraded in London, to the jubilation of the crowds, as 'a camelopard just arrived'.

When Mehemet Ali dispatched three orphaned young giraffes in 1827 to the three most influential heads of state in Europe, he was presenting them with more than just an imaginative gift; he was bestowing upon them the kudos of possessing one of the most mysterious and sought-after novelties that could ever appear in Europe – and at the same time sending the scientific world into paroxysms of excitement. 'The Year of the Giraffe' had begun.

That the animals arrived safely in Europe was nothing short

3

of miraculous, having endured the most tortuous journey, travelling strapped to the backs of camels from their place of capture in Senaar in Sudan to Cairo, before being buffeted about in ships' holds as they rode on rough seas from Alexandria to their various countries of destination.

The giraffe presented to King Charles X of France fared by far the best of the three, and was safely installed in the Jardin des Plantes, to the rapture of Paris. The animal sent to the Emperor of Austria, however, never fully recovered from the rigours of its journey. It survived only a few months in Vienna before dying of emaciation.

King George IV's giraffe arrived at its destination in relatively good health. But this was not to last long and the French, rather prematurely, began gloating over the demise of the 'Anglo-Camelopard' and boasting of the survival of their own 'unique' specimen in Paris. King George however was not to be deprived so easily of his prize. Long pilloried for his extravagant tastes and decadent life-style, he fell for the giraffe with characteristic passion. Every attempt was made to accommodate his new pet. A special building was erected for it in the King's menagerie at Windsor Castle, to which it was transported by regal procession in a 'spacious caravan'. There, in its own private paddock, the young giraffe was attended by Edward Cross of the Exeter 'Change, two Arab attendants, and the milch cows that had accompanied and fed it on its long sea journey.

For the poet Thomas Hood, the trappings of majesty with which it was greeted on its arrival on 11 August 1827, could not detract from the unhappy prospect of the poor animal's confinement. In his 'Ode to the Camelopard' he cries:

> Welcome to Freedom's birth-place – and a den!
> Great Anti-climax, hail!
> So very lofty in thy front – but then,
> So dwindling at the tail!

The King, however, became devoted to his giraffe, visiting it constantly in its quarters and tirelessly showing it to guests. He commissioned two portraits of it: the first by R. B. Davis, later animal painter to William IV, and the second, one of the King's last acts of patronage, by the great French painter Jacques-Laurent Agasse, for the princely sum of 200 guineas.

The King's almost immodest affection for the giraffe was well understood by at least one contemporary observer: 'Nothing could give an idea of the beauty in her eyes – Imagine something midway between the eye of the finest Arab horse and the loveliest southern girl, with long and coal-black lashes, and the most exquisite beaming expression of tenderness and softness, united to volcanic fire'.

The giraffe aroused similar, if less romantic, excitement in scientific circles: it was measured precisely from head to hoof as it grew; its spots were plotted; its teeth were counted; even its tongue was subject to a minute examination.

For others in Britain, the obsession with the young giraffe that year came to represent all that was reprehensible in the self-indulgent behaviour of the King. Caricatures lampooning the devotion of the King for his 'plaything' filled the papers. One, entitled 'The Camelopard, or a New Hobby', depicted the rotund figures of His Majesty and Lady Conyngham, his mistress, riding the delicate and startled creature around Windsor Park.

The giraffe, however, seemed never to have recovered from the rigours of its journey and in October 1829, despite every possible attention, it died. Fascination with the creature, though, did not diminish with its death. In keeping with the traditions of its native country, the royal pet was carefully mummified. It is at this point that John Gould's name appears for the first time in the press. *The Windsor and Eton Express*, for 17 October, 1829 reports 'Messrs. Gould and Tomkins of the Zoological Gardens, are now dissecting the Giraffe which expired on Sunday last. We understand that when the skin is stuffed, His Majesty intends making it a present to the Zoological Society.'

Gould's task was not an easy one. He had probably seen the

giraffe alive – an advantage he had over many taxidermists, who had never seen their subjects before they were skinned. The logistical problems, however, of setting up an animal nearly 11 feet tall with a neck four feet long were clearly enormous. Exact measurements had been taken of the animal while alive, and Gould and his colleague had to keep to these precisely. One false incision, one over-zealous stretching of the skin, and the model would be marred forever; a casual error in the process of drying and the famous prize would be subject to the ravages of mildew and mould. The result, however, of Gould's and Tomkins' exertions was an unmitigated success. At the age of 25 Gould had entered the scientific fray for the first time and proved himself a valuable asset to the British team. He had turned the disastrous demise of the British giraffe into a triumph of modern taxidermy.

Scientists in England revelled in the new-found opportunity of studying the creature to even greater exactitude: 'The economy of animals can only be studied when the functions of life are in full activity; their haunts must be explored, their operations watched, and their peculiarities observed in the air,' wrote the naturalist William Swainson in 1840. He continued, 'But in order to acquire a more accurate knowledge of their external form, and to investigate their internal structure, it is absolutely necessary to examine them in a dead state. Hence has arisen the art of taxidermy. . . .'

King George IV's giraffe received an even greater audience after its death than before.

It was a feat that Gould was to remain interested in, privately at least, for the rest of his life. When another giraffe came into the possession of the Zoological Society in the 1830s, Gould's secretary kept him informed of its progress. It too died prematurely and was stuffed in 1839, not by Gould but by the team of resident taxidermists then working for him. By this time Gould was free to devote himself entirely to his singular passion for birds.

CHAPTER 1

Stuffing Birds by
His Majesty's Order

GOULD HAD COME A LONG WAY by the time he found himself up to the elbows in a famous giraffe in the Museum of the Zoological Society in London. Unlike most of his colleagues, he came from a poor, working-class background, had little education, no public school or university training, and no connections in the aristocratic world of science.

Gould's father was one of ten children who came from a farming family in Somerset. He had moved away from his birthplace by the time he was 20 to take up a gardening job in the seaside town of Lyme in Dorset. There he married Elizabeth Clatworthy on 26 September 1803, and she bore him a son on 14 September the following year. For Gould these humble beginnings were more of an embarrassment than a blessing, although later commentaries on his life liked to glamorise them.

But it was his father's success as a gardener and his gradual promotion up the horticultural ladder that brought Gould closer to the pearly gates of the scientific world and placed him in a position to make his own enterprising move. As an occupation, gardening was coming into its own; it was one of the few professions with possibilities for the working man. The dawning of the industrial age, bringing with it a new class of prosperous and socially competitive entrepreneurs, had seen the blossoming of the suburban and cottage garden. The modern villa, built to accommodate the rising aspirations of the *nouveau riche*, began to emulate the grand landscaped gardens of the landed gentry on a miniature scale. The narrow streets and packed buildings of the old inner-city gave way to the avenues and vistas of sprawling suburbia. 'For a few decades', writes Geoffrey Taylor, 'before the full force of the Industrial Revolution had made itself felt, all England was a garden, and almost everyone in Great Britain was a gardener . . .'

There was also a partly aesthetic, partly scientific, interest developing in new and curious plants, many of which were being introduced from America and Asia. The London, later the Royal, Horticultural Society was founded in 1804, the year of Gould's birth, by the veteran botanical masters Thomas Andrew Knight and Sir Joseph Banks. Its aim was both to receive new plants from foreign lands and to send men out to find them. The next few years saw a vanguard of intrepid explorers setting out for the most remote corners of the world with the sole purpose of plant collecting. As a result thousands of different species were shipped to England in the hope they might be propagated at home.

Meanwhile, the nineteenth-century gardener's bible – Sowerby's *English Botany* – was in the throes of birth. Thirty-six plants, containing some 3,000 coloured illustrations, were produced over a 30-year period from 1790 to 1820. Within years the horticultural scene was swamped with garden journals and periodicals.

Garden tools and props, conservatories and greenhouses in particular, were being improved all the time; nursery gardens were established, and local 'Florists' Clubs' followed hot on the heels of the parent Horticultural Society in London. The status of 'gardener' was becoming increasingly prestigious; moreover, it was a position that lent itself to advancement. Paxton, Repton, and Capability Brown before them all rose through the ranks from humble labourers to leading lights via the propitious ladder of gardening. For the first time men and women were making names for themselves by virtue of their abilities and despite their social class.

Gardening was beginning to play a role too in the fortunes of the tiny seaport on the south coast. Lyme, which had not yet been regally dubbed, was, at the turn of the nineteenth century, a town on the make. Its new status on the social list – by virtue of a modish belief in the efficacy of sea-bathing – made it the object of substantial aggrandisement. On 14 September 1804, the very day of John Gould's birth, Jane Austen was writing to her sister from Lyme describing the social occupations of walking on the Cobb, playing cards, and dancing in the Assembly Rooms, as well as the snobbish competition between owners of the new villas in town.

The gardens of Lyme's villas were exhibited with singular ostentation. There was ample opportunity for a budding horticulturalist both to learn the latest gardening techniques and to display his skills to their best advantage. John Gould senior, as the Philpot Museum in Lyme Regis suggests, probably landed a job working for the Dowager Lady Poulett, whose house in Pound Street is now the Alexandra Hotel. Its grand and sophisticated glass-houses even included a special building exclusively for the propagation of myrtles.

Lyme Regis, with its attraction for visitors with social connections, was prime head-hunting ground. By the age of 23 Gould's father had been offered a better job on a large estate near Guildford in Surrey. The estate incorporated several acres of parkland, a large formal garden with ornamental trees and shrubs, a lawn, and a small lake; the house contained a large conservatory and had a kitchen garden. While there were great possibilities for a young gardener with ability and ambition, as in all the larger estates there was a strict staff hierarchy at Stoke Hall, and Gould's father must have set his foot initially very low upon the ladder. In the 1813 baptismal registration Gould's father is described merely as 'poor'. It was not until 1817, after 11 years of employment at Stoke Hill, that, on the baptism of his fifth child, John Gould senior was elevated to the rank of 'gardener'.

Of John's education nothing is known except that it was probably rudimentary. Gould was to despise this disadvantage, and later did all he could to conceal it by grandly philosophising on the character of a Latin name, including pretentious literary quotations in his ornithological texts, and disguising his inadequate grammar and spelling by depending upon an educated secretary for most of his correspondence.

Towards the end of his life Gould came closer to reconciling himself with his lack of education, and learned to play upon the romantic image of his being a child of nature. 'How pleasant it is,' he wrote in the letterpress of *Birds of Great Britain* (1862-1873), 'when the blessing of memory enables us to revert to some of the more agreeable events of youth, and almost of our infancy! How well do I remember the day when my father lifted me by the arms to look into the nest of the Hedgesparrow in a shrub in our garden! This first sight of its beautiful verditer-blue eggs has never been forgotten; from that moment I became enamoured with nature and her charming attributes; it was then I received an impulse which has not only never lost its influence, but which has gone on acquiring new force through a long life.'

In another Wordsworthian digression in the introduction to *Birds of Great Britain*, Gould elaborates on the almost spiritual influence of his childhood on his subsequent career:

'In writing upon subjects connected with ornithology I find the associations of my boyhood ever flitting before me. Well can I recollect the dried body of the brightly coloured Kingfisher hanging from the cottager's ceiling, and supposed by its movements to point the direction of the wind – a superstition now, like many others, happily abandoned. Well do I recollect also the particoloured strings of eggs with which I and my companions delighted to festoon the walls, and which were vigorously destroyed in our games before the termination of the year, in order to ward off the ill-luck otherwise supposed to ensue. I can still remember with what intense admiration I was filled in gazing upon the nest and lovely blue eggs of the common Hedge-Sparrow, and the pride I had in consigning them, when blown and thus bereft of half their beauty, to that string which was to hold so many of my subsequent findings . . .'

Later commentators, even during Gould's lifetime, emphasised an early predilection for studying birds. It helped to explain Gould's overriding obsession, and the extraordinary way in which he rose from obscurity to acclaim. It was typical of Victorian sentiment that Gould came to be considered as possessing a genius that was irrepressible. His aptitude was seen by most commentators on his life as a gift that would eventually have been widely recognised despite the difficulties of his background and upbringing. However, when Gould was just a boy there occurred a turn of events that was particularly fortunate in terms of his advancement, and set the young boy up in a position of spectacular advantage when it came to making a name for himself.

By the time Gould was 14, his father's gardening skills had attracted the attention of the court; and, as Lovell Reeve, who wrote a short account of Gould's early career in his *Portraits of Men of Eminence* in the 1860s, relates, his father, 'obtained an appointment as foreman in the Royal Gardens of Windsor under Mr. J. T. Aiton, and having removed his family thither, young Gould now commenced the active business of life in the royal gardens. "I've gathered many a bunch of dandelions," we have heard the ornithologist remark, "for Queen Charlotte's German salads."'

Gould was naturally expected to take up his father's profession, much to the amusement later of zoological colleagues. 'It was not in animated life that he found his first opening,' remarked one. 'He had to go through what some would think the severe ordeal of botany, and was occupied with plants and flowers.' Despite some glib comments that, as a result of Gould's apprenticeship to John Townsend Aiton, the head gardener at Windsor, 'a taste for botany and floriculture was added to his previous bent for zoology', the young Gould never seemed to embrace his prospects amid the borders and hot-houses of Windsor Castle: 'Botany is not my forte,' he once remarked in middle-age to a friend, despite the fact that in his last great publications he began to include plants and flowers as an attractive and informative background to his bird illustrations.

It may not have made a gardener out of him, but Gould's training at Windsor did introduce him to the art of taxidermy, a practice common among most gardeners of the time. The common garden birds and mammals stuffed and mounted 'after nature' were trophies treasured by the English gentleman on his country estate. We do not know who initiated Gould into the craft – again, little is disclosed about the sources of Gould's learning – but there must have been someone on the royal estate, perhaps Mr Aiton himself, who taught him the technique. From then on, every spare moment of the day was spent in search of specimens to stuff.

Taxidermy was still in its infancy, at least compared to the rage it became in mid-Victorian Britain, when Gould began. It was to be a while before every town had its own taxidermist, and almost every house its tableaux of stuffed birds or animals flaunting themselves under glass. Yet taxidermy as an art was still quite primitive. The advantages of arsenic as a preservative over the old recipes based upon sulphur and turpentine were only just beginning to be realised. For this reason few specimens of animals or birds stuffed before the 1830s have survived to the present day, apart from those treated by Gould. The British Museum possesses a pair of magpies shot and mounted by Gould in 1818, the year his family moved to Windsor.

Gould soon earned a local reputation, particularly among the 'young gentlemen of Eton College', where he began to sell his specimens for pocket-money. He received encouragement from the Rev. Joseph Goodall, then Provost of Eton, to whom he dedicated his first monograph on the toucans:

Rev. Sir,
    In dedicating to you this Monograph of one of the most beautiful and interesting tribes of birds, I wish to testify my feelings of respect and esteem for one, who, endowed with profound literary and scientific attainments, has long filled with so much ability the highest station in the most celebrated of our public colleges, that College being the pride and ornament of a town in which I resided for many years: and I am the more anxious to pay you this just tribute, as you have not only noticed and encouraged my own humble efforts to advance the study of that branch of natural history to which I have attached myself, but have ever most liberally contributed towards the promotion of science in general; permit me then to offer you this public testimonial of my respect and gratitude, and to remain,

    Reverend Sir,
        Your most obedient servant
        John Gould.

As part of his training as a gardener, Gould was sent when he was 18 to work for Sir William Ingleby at Ripley Castle in Yorkshire, where he was expected to 'learn the art of forcing plants' under the head-gardener Mr Legge. According to the pay-sheets in the Ripley Castle archives, Gould was employed between 18 September 1823 and 26 February 1825, and paid 12 or 13 shillings a week, depending on whether he worked six or seven days a week. Nothing else is known of Gould's time at Ripley Castle, except that it must have confirmed his preference for taxidermy over gardening. For, sometime around 1824, Gould moved to London to set himself up in the business. His Windsor connections stood him in good stead, for a receipt dated 1825 in the Windsor Castle Royal Archives indicates that he was entrusted with the unenviable task of 'preserving a thick knee'd Bustard' for King George. Gould is the first known taxidermist to have received royal patronage, although as taxidermy gathered momentum there followed a host of others endowed with a royal warrant to stuff things.

King George IV proved to be the perfect patron. Priding himself on being one of fashion's more outlandish pioneers, and fanatical about the collection of novelties he had brought together in his menagerie, His Majesty took to taxidermy like the proverbial duck to water. After the bustard Gould received a host of orders, from preserving 'two Mouse Deer, two Cotamundys' and an ostrich in 1826, to the final bill early in 1830 – only a matter of months before the King died – for stuffing the famous giraffe, together with 'a crane and two lemurs'.

Taxidermy was beginning to have significance beyond that of mere decorative art. As techniques improved it took on a role of scientific importance. For the first time specimens began to look like the original article; they became valuable objects of research, evidence of the natural world that could be scrutinised at a scientist's leisure. In the case of birds in particular, the most elusive of zoological subjects, taxidermy provided a tangible database for scientific reference. Bird-stuffing gave the fledgling pursuit of ornithology the impetus it needed to become a mainstream discipline.

In 1827, Gould participated in a competition at the newly formed Zoological Society of London. 'Gould's artistic talent in bird-stuffing, which he had studiously cultivated for some years,' recorded a notice in the Proceedings of the Royal Society, 'rendered him *facile princeps* among the competitors, and obtained him the appointment.' At the age of 23, Gould accordingly took up the post as first 'Curator & Preserver for the Museum' at the Society's lodgings in Bruton Street.

CHAPTER 2

Birds of a Feather

THE ZOOLOGICAL SOCIETY had only been in existence for a year when Gould was recruited as its taxidermist. Its arrival on to the British scientific scene, though, was long overdue: there had been no institution devoted to the study of zoology comparable to those on the Continent since the middle of the previous century. With the flood of new specimens of every sort arriving from all over the world – from minerals, plants, and shells, to animals, birds, reptiles, and even fish-skins – the country was sorely in need of specialised institutions to deal with the tidal wave in a more manageable and systematic way. The established institutions were swamped by contributions, and unable to raise their heads above the mess.

By the 1820s the vast zoological collections of the British Museum were mouldering away in 'an appalling state of chaos and neglect'. There was no public money granted for scientific research or the preservation of collections, and staff were paid a pittance for their efforts to control the mounting pile of miscellania. The naturalist William Swainson described the British Museum's basements as resembling 'the catacombs we have seen in Palermo, where one is opened every day in the year, merely to deposit fresh subjects for decay, and to ascertain how the process has gone on during the previous year.' The Royal and Linnean Societies still served a worthy scientific purpose, but their pioneering days were long gone, and they had become little more than unwieldy dinosaurs plodding along in a rapidly changing landscape.

The Zoological Society was the brainchild of Stamford Raffles, a remarkable man of irrepressible talent and ambition. The son of a ship's captain, he began life as a clerk at East India House, but had become by the age of 30 Lieutenant Governor of Java. He went on to become the founder of Singapore. But, quite apart from his political aspirations, he was also a fanatical botanist and zoologist, collecting as much of the local flora and fauna as he could from his various posts in the Far East, for identification and exhibition back in England.

As early as 1817 Raffles is said to have discussed with Sir Joseph Banks, the President of the Royal Society and founder of the London Horticultural Society, a plan for establishing a zoological collection in London. Raffles' interest was clearly stimulated by his visit that same year to the Jardin des Plantes in Paris, and his plans for an English version were modelled firmly upon it.

The galling example of the Jardin des Plantes flaunting itself just across the water in France sharpened the spirit of patriotism that was already making itself felt in scientific circles in Britain. Colonial competition was hotting up between the countries of Europe; the race to conquer and carve up the world had begun. What better way was there for a nation to demonstrate dominion over the dark and dangerous regions of the world than to bring back to civilisation trophies of strange and inscrutable wildlife? And how much more prestigious it would be to bring back these prisoners of empire alive, and then to divest them of all their fearful mystery under the powerful lamp of scientific inquiry for the greater glory of universal

17

knowledge. With missionary zeal, the prospectus for the proposed Zoological Society exhorts:

> 'Rome, at the period of her greatest splendour, brought savage monsters from every quarter of the world then known . . . It would well become Britain to offer another, and a very different series of exhibitions to the population of her metropolis; namely animals brought from every part of the globe to be applied either to some useful purpose, or as objects of scientific research . . .'

Expanding trade routes and a plethora of new colonies had put Britain in the perfect position to exploit the natural world; that she had failed as yet to realise the scientific, and hence political, advantages of her position was a matter of urgency to the founding father of the Zoological Society. Raffles determined to establish a scientific institution befitting 'the metropolis of the British Empire'.

On 29 April 1826 the 'Friends of the proposed Zoological Society' held their first general meeting in the Horticultural Society's rooms in Regent Street. Among the 48 people present was a radical nucleus of disenchanted Linnean Fellows, who had attempted to form their own Zoological Club within the Linnean Society. Long frustrated by the Linnean Society's bias in favour of botany, to the almost total exclusion of zoology, they had decided to redress the balance. But the constraints imposed upon them by their jealous and despotic parent – that membership be restricted to Linnean Fellows and Associates only, and that no paper containing original material be published without the Linnean Council's permission – had proved impossibly restricting, and the Club's members began to look elsewhere for an instrument for their purpose.

Foremost among the rebels were Joseph Sabine, also honorary secretary of the Horticultural Society, as well as a recognised authority on British birds; Nicholas Aylward

Vigors, a Member of Parliament and a fanatical zoologist with a special interest in birds; and Thomas Horsfield, an old friend of and collector for Sir Stamford Raffles, and keeper of the prestigious bird collection in the museum of the East India Company in Leadenhall Street. All three undertook important executive posts in the new Zoological Society at the first meeting: Sabine becoming its Treasurer, Vigors becoming Secretary, and Horsfield its Assistant Secretary.

Although the Zoological Club soldiered bravely on in ever-dwindling numbers, it bore the Zoological Society no malice. The developments of the new Society were reported regularly in the Club's *Zoological Journal*, and Vigors was congratulated on resigning as Club Secretary in order to take up his new post. Mr Bicheno, the Club's President in 1826, welcomed the Zoological Society with open arms, referring to it in his annual address as one which:

'. . . contemplates a more practical cultivation of science than any other which exists. They not only meditate the establishment of a museum, which has already been enriched by the private collection of Mr Vigors and the Sumatran collection of the late Sir Stamford Raffles [Raffles died unexpectedly less than three months after the first meeting of the Zoological Society] but every exertion will also be made to obtain an osteological collection, and in the end to establish a Menagerie, Aviary and Piscina. Every lover of Natural History will rejoice to hear that their Museum will be open to the public in the ensuing spring.'

The Club was finally dissolved on 29 November 1829, three years after the foundation of the Zoological Society. If the Club had not been able to keep afloat, it had at least provided a temporary life-raft for those of its members in search of a more permanent zoological haven. Perhaps the very justification for its existence lay in the stimulus it had injected into the fledgling Zoological Society. It was the core of ex-Zoological Club members that undoubtedly gave the Society its revolutionary flavour. Bent on transforming the stuffy and anachronistic approach to science that was typified by established institutions like the Linnean Society, they steamed ahead into the turbulent waters of controversy. While most other similar societies had remained little more than exclusive men's clubs, the Zoological Society took pains to promote itself as a genuine scientific establishment. Determined to prove its superiority over the institutions that had tried to suppress it, the Zoological Society proclaimed itself to be the true scientific standard-bearer of the British Empire. For the first time a scientific establishment threw its doors wide open, beyond the magic circle of the aristocracy, and welcomed the arrival of the professional scientist. Members were wanted, the Society declared, for their knowledge and ability, and not only for their influence and enthusiasm: 'In the first instance we look mainly to the country gentlemen for support, in point of numbers,' wrote Sir Stamford in 1825, 'but the character of the institution must of course depend on the proportion of men of science and sound principles which it contains.' In this spirit, the common practice of subjecting prospective members to an intelligence test was abolished; men with little formal education would not be impossibly handicapped. Finally, in an unprecedented step, women could be admitted. Admission depended only on the recommendation of three Members, and payment of the subscription fee.

It is questionable whether Gould would ever have attained such an important post as curator and preserver to a society that was older and less open-minded than the Zoological. It is certainly unlikely, however, that, had he managed to penetrate the hallowed halls of those more venerable institutions by merit of his taxidermical talents alone, he would ever have received the encouragement and vigorous promotion he found at the

Zoological Society. Gould was just the person they needed: obsessively hard-working, methodical, sharp, and religiously committed to the task they had assigned him. That he was uneducated and of humble background was of little concern to a society invested with the task of creating the world's definitive zoological museum.

It was the Museum more than anything that was to stand as a testimonial to the serious scientific objectives of the Society. At a time when science was grappling with the overwhelming task of putting a label to the mountains of newly discovered species, when it was still struggling to gain a foothold in the confusing world of natural history, having dead specimens preserved, named, and categorised side by side under the same roof was both necessary and profoundly comforting. Gould was charged with the sisyphean task of creating order out of chaos, something at which he proved extraordinarily capable.

When William MacGillivray, the hardy, eccentric Scottish naturalist, walked all the way from Aberdeen to London in 1833 to see the various societies and collections, he was impressed most of all by John Gould, that 'celebrated preparer of objects of natural history'. In his critical and sometimes fiery account of his visit, he was unreservedly appreciative of the display he found at the Zoological Society.

For those who were mistrustful of the menagerie of living animals being established in Regent's Park, the existence of the Zoological Museum vindicated the 'mere wild-beast-show part of the Society'. But sometimes, especially in the early years before enough cages had been built in the Park, the two worlds met, and the living species scampered, squawked and flapped disrespectfully around the skins of their fellows.

In the over-crowded nest of the new Society there was also hatching an interest in the science of ornithology, which had never before been regarded as a separate branch of zoology. Gould struck up a friendship with Nicholas Vigors, and together they set about putting ornithology on the map. Of his elevation from taxidermist to ornithologist, Gould's friend and colleague Bowdler Sharpe relates:

'My friend Mr Gerrard remembers him in these early days as a man of singular energy, with a good knowledge of the art of mounting animals, and indeed some of the best taxidermists in England were working under Gould at that time – such men as Baker, Gilbert, and others. At that time Vigors and Broderip were the moving spirits of the Zoological Society, then in its infancy, and from the former naturalist Gould received great encouragement. Vigors was an Irish gentleman and a member of Parliament for Co. Mayo. He was one of the most active of the early founders of the Zoological Society, and not only described several interesting collections of birds, but was the author of a "Classification" which exercised considerable influence on the minds of ornithologists for some years afterwards.'

Vigors was quick to recognise Gould's talents of observation. In his address at the last meeting of the Zoological Club, he stresses the valuable contributions made towards the 'elucidation' of ornithology in recent years:

'. . . above twenty species have been added to our catalogue since the date of the institution of our club, chiefly by the exertions of its members. The past year has not been barren of such acquisitions. A female specimen of a species of the family of warblers, the Sylvia Tithys of Linnaeus, shot near London, was exhibited at our last meeting [Gould's first scientific paper entitled 'On the occurrence of a new British Warbler' was published in Volume V of the *Zoological Journal* in 1835.] We are indebted to Mr. Gould of the Zoological Society, for the discovery of this addition to our British Fauna. The bird had been sent to him as a Common Redstart (Sylvia Phoenicurus), to which it bears a close affinity; and probably would have passed unnoticed as a specimen of that species, most particularly in consequence of its sex, in which the colours are less strongly marked than in the male, had not the critical knowledge of this rising naturalist detected the distinguishing characters.'

Gould had found in the Zoological Society a mentor as passionate about birds as himself; he had also discovered a golden key to the scientific world and an introduction to the men of influence who presided over it. United in their aims and common interest, the members of the Zoological Society would do all in their power to help the progress of a colleague. Gould's position was one that the great American bird illustrator John James Audubon bitterly envied. Resentful of the closed circles of the British aristocracy and yet desperate to be accepted by them, Audubon had endured a frustrating and sometimes humiliating year after his arrival in England in 1827 as he attempted to interest various people in the publication of his work. In 1835 he commented to his great friend J. Bachman that Gould was a man to watch: 'Gould is a man of great industry, has the advantage of the Zoological Society's Museums, Gardens, &c – and is in correspondence with Temminck, Jardine, Selby, James Wilson and the rest of the Scientific Gentry.'

By 1833 Vigors had recognised both the potential of a separate division of ornithology within the Society, and the ability of the Museum's curator to preside over it. He appointed Gould superintendent of the ornithological department, but by then Gould had published his first great volume of bird illustrations, and had embarked on another. Both the advance of ornithology and Gould's career were already well under way.

At the time of Gould's first appointment to the Zoological

Society a number of major ornithological publications were in progress. It was a time of frenetic activity in the field of bird illustration. Britain had just broken the spell of torpor that had reigned almost uninterrupted since the publication of Willoughby & Ray's *Ornithologia* as far back as 1676.

While the *Ornithologia* was nothing short of a landmark in the history of the science, it had revealed a number of problems concerning the art of bird illustration. There was clearly a scientific need for pictures to describe and identify different birds, just as there was a need for illustration in botany or entomology. The subject matter, however, could be maddeningly elusive: if it could not be captured, it had to be shot, which usually meant blasting the creature beyond recognition and rendering it little more than a sorry bundle of feathers. Drawings made from reconstructed corpses crudely preserved, or stuffed and mounted, were little more than representations of rigor mortis. There were problems, too, in the printing of drawings. Most gentlemen could draw, particularly if they were naturalists, but engraving required a professional, who was sometimes woefully ignorant of the nature of the subject. In his preface to the *Ornithologia*, Ray laments not only the crippling expense of hiring engravers, but also their inadequacy in depicting birds.

While the world of bird illustrations in Britain was cowed into a respectful silence for well over a century following these realisations, production on the Continent was not so inhibited. The middle of the eighteenth century saw the rise of the great French naturalist Le Comte de Buffon and the publication of his epic 44-volume *Histoire naturelle, generale et particulière* – begun in 1749 and finished after his death in 1788 by Lacepede, and the triumphant *Planches enluminées* (1765-1786). Buffon's works, generously furnished with elegant engravings, were translated into several languages and swiftly earned acclaim throughout Europe and North America. Like most books of the era, it set its sights on a broad horizon, embracing the entire spectrum of natural history from the animal to the plant kingdom, without focusing on any particular aspect. Of the 1,008 plates included in *Planches enluminées*, however, most were of birds.

Buffon's countryman, Françoise Levaillant, was one of the first to limit his scope, and from 1796 to 1808, not long after the great master's death, he produced his exquisite *Histoire naturelle des oiseaux d'Afrique*, with coloured plates. He also produced between 1801 and 1805 one of the first works dedicated to one particular species, his beautiful *Histoire naturelle des perroquets*.

At the same time, the Scots-born Alexander Wilson was staking similar territory in the New World. His *American Ornithology*, begun in the year of Gould's birth, was subscribed to by, among others, Thomas Jefferson. It describes 264 species of the 343 birds found on the continent at the time, with no less than 48 of these being new species discovered by Wilson himself.

But it was the French-born American, John James Audubon, who took the art of bird illustration to its most extravagant and flamboyant extreme when he began to produce in the 1820s the largest engravings ever before published. Produced on massive double-elephant folio paper (*c*.27 × 40in/68 × 101cm), Audubon's *Birds of America* are gorgeously coloured, mostly life-size, expertly drawn, and have a sense of freedom and drama never before seen in zoological illustration. The popular appeal of his illustrations greatly influenced the scale of many future publications, although no one ever quite dared to try to measure up to the American.

The commotion on the two continents washed also against the stolid isles of Great Britain, and awakened her at last from her dry and relatively unproductive century. It was Thomas Bewick, a fervent naturalist with a training as a Newcastle engraver, who revolutionised the art of wood engraving in his attempts to depict the scenes of nature that inspired him as faithfully as possible. Bewick was in the privileged and unique position of being both artist, engraver, and naturalist. He was passionate and knowledgeable about his subject, and infinitely capable of reproducing it. In publishing his influential work *British Birds*, Bewick's object was 'to stick to nature as closely as he could'. For the first time in Britain birds were depicted in their natural habitat and as if they were really alive. 'Elegant and accurate figures,' urged Bewick of his fellow countrymen, 'do much illustrate and facilitate the understanding of descriptions.'

For many, Bewick's woodcuts were a revelation. 'I have heard those who loved the country', wrote William Howitt in the *Rural Life of England* in 1844, 'and loved it because they knew it, say that the opening of Bewick was a new era in their lives.' The unassuming northern naturalist was a major influence on the early interests of Gould while he was still employed as a gardener at Windsor Castle. For Audubon too the works of Bewick were a landmark in the development of bird illustration, and he visited the artist at home in his workshop when he came to England in 1827.

Bewick was the first illustrator of birds, at home or abroad, to make an issue out of drawing from life. Although an excellent shot, Bewick hung up his gun forever when he was still a young man. While he accepted skins from other ornithologists and collectors, and, when he was not able to study the bird alive, worked from them for his book, he preferred to rely on his telescope, and an inordinate amount of birdseed, to master his subject. This insistence, together with a determination to be as faithful to nature as the pen or the stylus would allow, became Bewick's legacy to the succeeding generation of ornithological illustrators. In his preface to *British Birds*, Bewick remarked of the earlier works of Buffon, 'that in many instances that ingenious philosopher has overstepped the bounds of Nature . . . too frequently hurried into the wild paths of conjecture and romance.' The sentiment set a trend for future illustration in Britain – it was the very criticism, indeed, that Gould was later to level at Audubon – and it stimulated a keen sense of competition between the British and their contemporaries on the Continent, such as Temminck, Vieillot, De Blainville and L'Herminier. Britain was to make up for its tardy entrance on to the ornithological scene by producing more books on birds in the nineteenth century than all of the other European countries put together.

For the young curator and preserver of the Zoological Society, bent on creating a name for himself as an ornithologist, entering the fray of the publishing world would have seemed inevitable and eminently desirable. It was the one way in which Gould could establish himself once and for all among the ranks of the scientific fraternity where he had gained only a foothold as taxidermist. But bringing out an illustrated book off one's own bat was a speculative, daring, and expensive step to take – especially for someone like Gould, who had no training and little natural aptitude in drawing, no experience of publishing, and no real grounding in ornithology. His lack of formal education was criticised by Alfred Newton, author of the famous *Dictionary of Birds* and professor of zoology and comparative anatomy at Cambridge, as late as 1864. In a letter to his brother Edward, Newton remarks: 'By the way Gould always sends me his *Birds of Great Britain* to look over for him, and the utter ignorance they sometimes betray is amazing. He has no personal knowledge of any English birds, except those found between Eton and Maidenhead, and about those species he fancies no one else knows anything.'

Gould did, however, have the support of eminent and learned figures at the Zoological Society, many of whom were now producing books themselves. Prideaux John Selby had begun publishing *Illustrations of British Ornithology* in 1821, the first-ever attempt to produce a set of life-size illustrations of British birds. This was not completed until 1834, but Selby was also publishing, with his great friend and colleague Sir William Jardine, four volumes entitled *Illustrations of Ornithology* (1825-43). Selby and Jardine were to become intimate friends of John Gould, invaluable allies in the competitive search for new species, and early subscribers to his publications. It was for them that Gould produced his very first plates, which were published in *Illustrations of Ornithology*. Jardine in particular gave Gould encouragement in the early days, and became his first serious patron.

Gould's skill at stuffing things had landed him at the epicentre of a vigorous new scientific movement. That he was able to work side by side with men like Vigors, Broderip, Selby, and Jardine, the pioneers of ornithology, at a time when the science was ready to promote anyone who could be of advantage to it, was encouraging. But it was the way in which Gould was able to exploit his position to the full which got him so far, so fast. Less than four years after his appointment as curator, Gould was publishing his first book, on a level-footing with the gentlemen scientists around him who were doing the same. He had lost no time in building up the network of friendships and working relationships that enabled him to do this. 'John Gould', wrote Alfred Gunther after meeting him for the first time at the Zoological Society, 'was bon-ami with everyone who could be of any use to him.' Members of the Zoological Society provided him with the crucial number of subscriptions he needed to ensure the financial success of the undertaking, but most important of all, they provided him with their own example. The Fellows gave Gould the stimulus as well as the means of achieving his fame and fortune.

Despite taking a financially independent course early in his career, Gould was to preserve close ties with the Society for the rest of his life, producing over 300 papers at its meetings, acting as corresponding member during his travels to Australia, and twice becoming vice-president. Gould's progress through the ranks of the Zoological Society was nothing less than remarkable. He became, as the *Proceedings of the Royal Society* pointed out, 'an active Member of Council and Vice-President of the Society with which he had become originally connected in the humble capacity of bird-stuffer.'

CHAPTER 3

The Start of a Name

I N PURELY LOGISTICAL TERMS Gould was ill-
equipped to publish an illustrated work by himself. He
had a sound and instinctive business sense – a quality that
was to stand him in excellent financial stead for the rest of his
life – but he was unlettered compared with his colleagues, and
was no draughtsman. With typical panache, however, he solved
both problems by marrying a competent artist in January 1829,
and by securing the services sometime in the following year of
a dextrous and dedicated secretary. Both surrendered themselves completely to the great designs of John Gould. Each
died, Mrs Gould after 11 years, Edwin Prince after a service of
45, with their backs against the wheel.

Gould's debt to his two principal collaborators was immense,
at least to those who were granted an insight into the mechanics
of the Gould machine. 'To his wife's zeal and self-sacrificing
devotion', says the biographer of an artist who later also worked
for Gould, 'it is evident that he owed much of his after success.
His own industry, enthusiasm, and perseverance were beyond
praise; but without this timely help, and the ceaseless labour of
a long-suffering slave, one Prince (called by courtesy a

'Secretary') the result might have been different.' Prince's contribution to the laborious and complicated task of administration was appropriately acknowledged by Gould in later life. But the part his wife took in his productions was thought by some to have been underplayed. The coincidence of Gould's marriage with the appearance of his first publication the following year was not lost on his commentators. So quick was Gould to capitalise on his wife's talents that some considered Gould's publishing ambitions were the prime catalyst behind the marriage.

Little is known about Elizabeth Coxen before she married, except that she was one of nine children born in Ramsgate, and her father, Nicholas, is thought to have been a captain in the merchant navy. Elizabeth was, however, an educated young woman, with experience as the governess of a prosperous family living in St James' Street. In the only surviving letter written before her marriage, she writes to her mother describing her position teaching Latin, French and music to a nine-year-old girl. She confesses that she occasionally feels 'miserably – wretchedly dull' and fears that she 'shall get very melancholy here'. According to Gould family legend the couple met in the Zoological Society where Elizabeth had one day taken her charge, and fell in love 'at first sight'.

It is perhaps more realistic, if less romantic, to consider that for both this was a marriage of convenience, and that the match was more likely to have been made not in the gardens of Regent's Park or some grand country seat, but amid the wheelings and dealings of Broad (now Broadwick) Street in Soho; that it was not so much the collusion of Cupid and Aphrodite as the allurement of camphor and arsenic that brought Gould and Elizabeth together. For the most obvious connection between the two was another taxidermist, a Mr Coxen, Birdstuffer, of 12 Broad Street, Golden Square, a colleague of Gould's and a close relation (perhaps even a brother) of Elizabeth's.

Gould had already become acquainted with Broad Street by 1826, the year before his election to the Zoological Society, when he sent a bill to Windsor Castle giving 11 Broad Street, Golden Square as his address, and he is likely to have known Coxen since then. A contact in the world of taxidermy would have been a great advantage, but the area around Soho was an important focus for this kind of work. One of the most successful dealers in skins who catered for a clientèle of serious naturalists, Leadbeater & Son, was based just around the corner in Brewer Street.

The Broad Street connection was equally important for Elizabeth Coxen, who almost certainly met her future husband on a visit to her Coxen relative. But we can only guess if it was to Gould that she referred when she told her mother why she refused a position at Stratford-upon-Avon in the autumn of 1827. 'I certainly should have been a little lonely there,' she said, 'my friends remain dearest to me in Broad Street.'

By 18 March 1830, a year after their marriage, the couple had moved in with their go-between at his house at No. 12. And by 1 September of the same year they had moved further down the road to their own house at No. 20, where Gould was to remain for the next 29 years.

Whether he had chosen dispassionately or not, Gould could not have picked a more appropriate wife than Elizabeth Coxen. It was a great practical advantage for Gould to marry within the ornithological clan, and one that would not have been out of keeping with the times. Within the relatively insular world of natural history, a man like Gould was likely to gravitate towards a wife with scientific connections, if only because he had ample opportunity to meet her through the same society or club, at one of the numerous scientific meetings or social soirées, or through his work. But it was often a conscious decision on the part of the naturalist to marry someone who not only had an affinity for the work already, but who might also have useful connections and who would find the eccentric and demanding life-style of the naturalist husband a home from home. A wife who could contribute to a naturalist's work was also a considerable economy as she could hardly expect to be paid.

Gould had found in this young and frustrated governess more than just an assistant to help him stuff things for the Museum in Bruton Street, although she probably helped him with this task too. He had discovered the perfect partner to fulfil his publishing ambitions: Elizabeth was determined, intelligent, educated, practical, obedient, and she possessed the one attribute he most desperately lacked – she could draw.

Although almost all of the 2,999 plates that constitute the gigantic works of Gould were taken from sketches made by Gould himself, not one is known to have been executed by its 'author'. Gould's genius lay in his understanding of birds, not in his ability with a stylus or a pencil. While he could resuscitate in his mind's eye the dried skin of a bird on the table in front of him, he could not repeat the process on paper. His drawings mark out only the shape, size, and posture of the bird in very basic form. While Gould could never describe the quality of feathers or the brightness of eye, he could indicate something of the characteristics of his subject by marking out its most natural pose, and setting it against what he knew to be its natural habitat.

When laid alongside the drawings of Wolf, Lear, Richter, or Hart, as they are in the bound originals at Knowsley Hall, Gould's sketches stick out like a sore thumb. Some of them are childlike in their simplicity and awkwardness. Outlines are often fudged or over-emphasised with broken strokes. The lead of Gould's over-worked pencil gleams out from murky blotches of badly mixed watercolours. Browns and dirty greens dominate, so that often a bird merges with its perch, or loses its feet in the ground. Sometimes the sky is smudged with a sickly mixture of pink, purple, orange, and yellow pastel chalks – a child's parody of a sunset. Occasionally a bird carries something messy in its beak, which only later is recognisable, once it has been artfully translated by the lithographer, as a beetle, fly or worm. Certainly these were, as Gould so often emphasised, 'very rough sketches', drawn in 'great haste', but they have none of the confidence or dexterity of a casual sketch by any of Gould's legitimate artists.

His inadequacy as a draughtsman was a well-kept secret. He almost certainly intended his subscribes, his fellow scientists, and the general public to believe that he was as much an ornithological artist as a Bewick, a Selby, or a Swainson. He often autographed drawings that were patently not his; he labelled his plates with an ambiguous yet suggestive, 'Drawn from nature & on stone by E & J Gould', or '[artist's name] & Gould del et lith'; towards the end of his life he was caught pretending he was at work on a drawing that was actually someone else's. The living legend of Gould – the publishing magnate, ornithological genius, entrepreneur, producer, and director – was well founded; but Gould the artist was a self-perpetuated myth. Perhaps the wily young man realised even at this early stage that it was this last attribute that was most likely to immortalise his name.

Gould's artistic attempts, however, were far from superfluous or irrelevant to the plates that made him famous. No matter how displeasing they may seem, Gould's drawings were an integral part of the production; despite their amateurishness, they convey a significant impression of form and substance. It is easy to see from the finished plate how much the picture relies on Gould's initial conception; it is not so easy, in reverse, to imagine the finished plate from the original Gould sketch. This is where the genius of Mrs Gould, and subsequent well-chosen lithographers, becomes significant. With almost telepathic instinct they learned to read Gould's sketches like a template, and to interpret what he wished to portray. From Gould's first clumsy draught the lithographer was expected to make a detailed drawing, which would then be corrected by the boss. Certainly this must have been frustrating on occasion. In the early years before Gould made his name it is difficult to imagine how anyone, aside from Gould's infinitely tolerant and accommodating wife, would have been able to endure this laborious process, compounded by the demands of a dedicated taskmaster such as Gould.

Gould's brilliance was as a producer. His skill lay in his direction and criticism of the drawings of others. He understood the nature of birds as few other men ever have. With precision and confidence he would mark a fault in the shape of a beak or the angle of a leg, change a gesture or the tilt of the head, annotate a comment on the colour of the breast until the picture became what Gould saw in his mind's eye.

Gould was a considerable motivating force. He conceptualised his pictures even if he could not complete them himself, and he inspired and instructed his wife with fanatical enthusiasm. By September 1830 he was contributing her drawings to Selby and Jardine's *Illustrations of Ornithology*, then in progress. In a letter to Sir William Jardine he writes:

Sir,

I take the opportunity of forwarding 3 Drawings, which are the only ones I have been able to get done for you owing to Mrs Gould being very much indisposed.

She is now much better and Mr Vigors has this day looked out several birds to be figured by you, which shall be done immediately and forwarded.

The reason I did not send your Heron before this was my not being able to send the drawings, but you may now expect them soon with some other Birds that I have collected for you.

I remain Sir,
Your Obedient Servant
John Gould.

Soon after, Gould was given the golden opportunity to publish under his own name. One of the first major acquisitions of the Zoological Society arrived in Bruton Street towards the end of 1830. It consisted of an unprecedented 'collection of bird skins from the Himalayas – a district at that time almost unexplored'. It was a coup for the fledgling institution, which could now boast responsibility for over 25 new species. *The Times* commented, after the ornithologist's death, on Gould's significant decision to publish an illustrated work on the new prize from the Himalayas: 'In 1830 the accident, as it would seem, of his acquiring a fine series of birds from the hill countries in India determined his life's work in that direction. The speedy result was his *Century of Birds from the Himalayas*.'

So Mrs Gould, now pregnant for the second time (their first child, born exactly nine months after their marriage, had died) was set up as the resident company artist at their house in Broad Street. The work of an illustrator should not be under-estimated; it could prove a strain to even the strongest and most accomplished of artists. And for a novice, under the pressures of a strict timetable, and heavily pregnant, the work must have been doubly demanding.

By 30 November word was getting around in the scientific circles of the imminent publication. 'Gould is about to publish Illustrations of Ornithology from the Himalaya Mountains', writes Selby to his co-author Jardine, 'he has received an importation of very valuable skins from that interesting part of Asia, the majority of them quite new, the figures are of the natural size drawn & lithographed by Mrs E Gould, & the letter press is to be furnished by Vigors. The specimen he sent me is very tolerable . . .'

On 21 December 1830, John Henry Gould was born, less than a few days after the more celebrated nativity of *A century of birds hitherto unfigured from the Himalaya Mountains*. The day before the birth of his son, Gould wrote to Jardine, who had expressed an interest following Selby's letter, proudly enclosing copies of his first, cherished creation:

20 Broad Street, Golden Square
London Decr 20, 1830

I have this day forwarded . . . the first no. of my new work on the birds of the Himalaya Mountains which I hope will meet with your approbation and support. You have kindly said in your letter that you would endeavour to procure some subscribers for me for which I beg to add my thanks. I have sent you 2 nos, one I hope you will think proper to retain for yourself, the other you will be pleased to do as you think proper with and if you can procure me a few subscribers you will be conferring on me a great obligation. I am very well supported by the scientific gentlemen of London etc having allready [sic] 50 subscribers. Many persons say it is much to [sic] cheap but if I am as liberally supported in other parts as in London I shall have no reason to complain. I shall limit my number of coppys [sic] to about 200 and then destroy the stones, therefore every coppy [sic] sold will make the work more scarse [sic] and consequently more sought after. From the support I have allready [sic] received I do not fear selling the whole before the work is completed, which will handsomely repay me. If it is not taking too great a liberty will you be kind enough on the receipt of the box [to] answer this letter saying how you approve of the first number and any suggestions you may think proper to make.

I should be pleased and obliged for the trouble you had taken. It is Mrs. Goulds [sic] very first attempt at stone drawing which I hope you will take into consideration.

The skins of birds I have sent are the best of the kind I could procure which I hope will be of service to you, in return for which I should be glad to receive at any time a

fine pair of Merlins or the skin of a Black Throated Diver.
     I remain sir
         Your very obedient Servant
           · John Gould
P.S. As your name will add considerable support to my
work I hope I shall not be long before I have permission
from you to place it among my subscribers.

Mrs Gould was given a month to recover from the birth of her
son before the next part was scheduled in February 1831, with
a further part expected monthly thereafter. 'Have you got
Gould's nos 4 & 5 of Himalayan Birds,' wrote Selby to Jardine
on 26 April, 'they contain some good things & are very well
done upon the whole (they came to me yesterday night) she has
however in all the figures of the Pheasants made a serious
mistake in the foot, having placed the Scallop upon a level with
the other toes. I like them as well as Audubons . . .' It was an
inevitable comparison given the current nationalistic sense of
competition, and one of which Gould himself was well aware.

Robert Havell, the London engraver, wrote to Audubon and
commented: 'Mr Gould has been very successful in his [book]
of Birds [of the Himalaya Mountains], which has caused him to
think not a little of himself.'

Part 20, the final part of the *Century*, appeared shortly before
17 April 1832. Gould's schedule of four plates a month had
been a demanding one. That Mrs Gould was expected to
produce on average one finished plate a week is a mark not only
of her husband's faith in her ability, but of her own determination; that the entire work – 80 plates figuring 102 birds
(the two extra birds are chicks, a charming detail that had never
before been included in ornithological illustration) – was
finished five months ahead of schedule, is extraordinary. Gould
must have been pleased with his young wife's labours although
we hear nothing from him about it. At least Vigors made known
his appreciation of Mrs Gould's work by naming one of his

prize new species *Cinnyris Gouldiae* (Plate No. 56) after 'Mrs Gould, who executed the plates of these Himalayan birds'.

Vigors' own part in the production, however, was not proving as satisfactory. While Gould was frantically churning out illustrations, his colleague was disappointingly unforthcoming with the accompanying text. As late as 5 November 1833, a year-and-a-half after the completion of the plates, Gould wrote to Jardine, barely concealing his irritation with Vigors: 'I have forwarded in a parcel to Mr Wilson of Canaan the Letterpress descriptions to my work on the "Birds of the Himalayan Mountains", which will enable you to have your copy bound. I regret that it has been so long in hand which must be attributed to Mr. Vigors' political career which, I am sorry to say has taken him almost entirely from the pursuit of Natural History.'

Never again would Gould allow himself to be compromised by someone else's priorities. Quite apart from risking the dissatisfaction of his subscribers, whose allegiance could be hijacked all too easily, he had laid himself open to attack by rival zoologists. The Himalayan birds were too tempting a prize to be allowed to wait before being formally claimed by the printed word. Jardine wrote to Selby in November 1831, referring to two potential predators circling the catch at the Zoological Society:

Jardine Hall

My Dear Sir,

By a parcel from London I received the enclosed numbers of General Hardwicke's Indian Zoology and a new Zool. Miscellany commenced by Mr. Grey. Mr. Grey is a most indefatigable person with his miscellanies, [sic] Synopsis's & Spicilegice &c &c, but upon my word none almost of his species (of birds at least) are to be depended on. In Genr Hardwickes Numbers, some of Mr. Goulds Birds are figured & I suppose we shall next have a Quarrel about priority – Gray who appears to have been much about the Zool. Soc. lately must have been aware of the Himalayan birds, & it looks curious that both should have been going on as it were to get the start of a name . . .

If Hardwicke and Gray had behaved unscrupulously by attempting to cash in on someone else's prize, there are equally suspicious circumstances surrounding Gould's claim to at least part of his collection of the Himalayan birds. No one knows who was responsible for the original donation to the Zoological Society, and there are no known objections to their use by Gould in his publication. But in 1832 there came into the Society's hands, around the time that the *Century* was completed, another collection of birds from Nepal belonging to a Mr B.H. Hodgson, who was then serving in the British army in India.

Hodgson aspired to publish a work on the animals and birds of 'Nepaul', and was dismayed to find that not only did some of the species for which he had assumed responsibility figure in Gould's *Century*, but that Gould was preparing a follow-up publication that would undoubtedly eclipse any other work on the subject. Hodgson felt sure that Gould was taking advantage of the collection that he had deposited with the Zoological Society. Like so many other naturalists, however, interested in seeing their name attached to a prestigious, new publication, Hodgson lacked the financial and scientific backing, as well as the artists, to carry out the work successfully himself. It seemed more prudent to propose a collaboration than to incur the wrath of a man in such a strong position as Gould.

Gould's reply to Hodgson's proposition was long in coming. When it finally arrived on 6 March 1837, with scant apologies, it contained a proposal for a joint venture – on Gould's terms and lightly dusted in a coating of complacency. In his letter Gould insinuated that while Hodgson's collection may be of 'the highest interest not only to the scientific public but to an extensive class of the Community at large', it was by no means the only source upon which Gould could draw for publication:

> 'I beg to propose a plan for publication in which I conceive our joint labours might be incorporated and thereby produce a work rivalling any hitherto attempted. I am aware of what Mr. Hodgsons collections consist, at least those forwarded to the Zoological Society (which are of course kept sacred) and on my own part I could add considerably to the number of species having, since the completion of my 'Century' characterized many from [my] own collection which I had not observed in Mr. Hodgson's and having free access to the specimens contained in the collections of Fort Pitt, Chatham, the United Service Museum and nearly all the other Museums of this country as well as several on the continent . . .'

With the ease of a master negotiator, Gould deftly played his trump cards: money – 'I shall bear the whole expense of the publication on condition that Mr. Hodgson shall take 100 copies . . . the work shall remain my property . . .' – and reputation – '. . . as many of my Subscribers in England would readily take a continuation of that publication [the *Century*] while they might not [be] so ready [to] subscribe to so large a separate undertaking . . .'

Gould had Hodgson over a barrel. As Hodgson's father pointed out in a set of remarks sent to his son in reference to the proposals, 'Whatever J.G.'s reasons for waiting – he is losing no ground – his resources from various Quarters are multiplying surely and rapidly and his preparations would be available whether for a joint work or a separate one.' To Gould's suggestion that the work 'be published under the joint names of Gould and Hodgson as is the case with the "Planches Coloriées" of Temminck & Langier, the "Illustrations of Ornithology" of Jardine and Selby etc. etc.', Hodgson's father retorted angrily, 'Gould & Co. – & hardly Gould & *Co* – for at this rate what is it but a continuation of J.G. solus however other initials may be suffered, perhaps here & there –' Hodgson's father summed up Gould's proposals as nothing short of a con: 'For 1 or 1½ years B.H.H. to do nothing but send materials after materials – 'his manuscripts & other matters' – neither knowing what use will be made of them – or whether any or none – losing ground all the time – while J.G. is gaining, at any rate by keeping B.H.H. out of the field – profiting by time to forward the work of Engravings (for whatever use) and by the constant accession of fresh subjects.'

In a last-ditch attempt to persuade Gould to give his son greater concessions, Hodgson's father wrote directly to Gould from Canterbury on 10 March 1837: '. . . Surely it is too much to expect him so far to resign the credit of a work essentially his – as to take a *secondary* part in it – had he not distinctly so expressed his feelings – in his Letter – amongst others – to Mr Vigors – of May 1835 – within your reach I conclude as a Document of the Society . . .' Gould, Hodgson's father maintained, had taken unfair advantage of his position at the Zoological Society by intercepting his son's plans and contriving instead 'to lead all the world to regard an intended New Work but as a continuation of [his] former publication.'

The second publication on Himalayan birds, whether as an independent work by Gould or as a collaboration, was not to be. There were, it is true, extreme differences between Gould and Hodgson's ideas, quite apart from the conflict of egos. Hodgson wanted a publication on both mammals and birds; Gould's interest was purely in the birds. 'As Ornithology alone is my particular study,' he states categorically, 'it is this department only of the zoology of Nepaul that I would in any way engage to undertake.' It is difficult to believe, however, that Gould simply backed down over the issue. His determination was such that if he had really wanted to undertake another Himalayan publication, he would surely have found ways to reconcile Hodgson and his recalcitrant father. Gould seems instead to have simply discarded the idea, as he did with several other projects, through sheer weight of work. Immediately after publication of the *Century*, flushed with its incredible success, Gould had launched himself into the world of publishing, bringing out in quick succession a series of new works, one of which, a motley volume entitled *Icones Avium, or Figures and Descriptions of New and Interesting Species of Birds from Various Parts of the World*, gave him ample scope for figuring any new species from the Himalayas that came to hand. It soon became clear that there was no need to publish a sequel; Gould had moved on to greater things.

The *Century of Birds from the Himalaya Mountains* had served its purpose: it had made a name for John Gould. It had also earned him a reputation for being ruthless, if not downright unscrupulous, in his eagerness to secure 'new and hitherto unfigured species'. Gould had been remarkably successful in establishing the nucleus of his company. Prince was to become an indispensable component of his business (which still included taxidermy, and came to incorporate the buying and selling of bird and animal skins, and even the trading of other zoological works); while Mrs Gould was his ticket to illustrative publishing. The subject matter for his first publication fell with remarkable good fortune right into his lap; and members of the Zoological Society provided him with crucial business contacts and possible subscribers. There was another influence, however, perhaps the most significant of all, upon which the success of Gould's first enterprise depended. He might never have dared embark upon the *Century* had it not been for the example and instruction of a strange young illustrator working alongside him at the Zoological Society.

CHAPTER 4

Lear's Years of Misery

EDWARD LEAR, famous for his watercolour land-
scapes and nonsense rhymes, began his career at the age
of 16 under a cloud of poverty and insecurity as a
draughtsman to the Zoological Society. With a hopeless head
for finance and a desperate desire to befriend, Lear was only too
willing to receive the attentions of John Gould, then in his mid-
20s, and to initiate him into the secrets of the new technique of
lithography and the art of ornithological illustration. It was
Lear's example that provided the impetus for the Goulds' first
publication, and it was Lear who later transformed Gould's
static and rather unimaginative style into the confident and
innovative work that characterised his second and all subsequent publications.

Towards the end of the century, however, when Edward
Lear was living out his last years in Italy, a sick, embittered,
and lonely old man, he was to regret his early generosity, and to
resent the relationship that had evolved from it. On hearing of
Gould's death, Lear remembered, 'He was one I never liked
really, for in spite of a certain jollity or bonhomie, he was a
harsh and violent man. At the Zoological S. at 33 Bruton St. –

at Hullmandels – at Broad St. ever the same persevering hardworking toiler in his own (ornithological) line, – but ever as unfeeling for those about him. In the earliest phase of his bird-drawing he owed everything to his excellent wife, & to myself, – without whose help in drawing he had done nothing.'

Edward Lear was born in 1812, the penultimate of a staggering 21 siblings, at Highgate, North London. His mother surrendered his upbringing to a sister, Ann, 22 years Edward's senior. Lear was always to suffer from this apparent rejection. What little is known of his early life is cloaked in melodrama and make-believe, but it appears that his family, once prosperous but fallen on hard times in Lear's early years, sold up and went their separate ways. Lear moved with Ann to lodgings in Upper North Street, off the Gray's Inn Road.

Lear was a lonely, awkward boy, frightened by depression, the deaths of four of his sisters, and frequent attacks of epilepsy, which he referred to as 'The Morbids' and 'The Terrible Demon'. He found relief from his fears and inadequacies in a world of fantasy and art. In particular he made a study of natural history, especially of birds, with which he felt a strange affinity. Many of his later nonsense rhymes, concocted for the amusement of his patron's, Lord Derby's, grandchildren, owed their inspiration to Lear's fantastical obsession with birds, and their similarity, both in behaviour and appearance, to people he knew.

Lear was educated by his sisters, Ann and Sarah, both talented artists, who also taught him to draw..He later recalled the days spent in the 'painting room' immersed in his father's collection of books and pictures as the happiest of his life. He copied, among others, all the animal plates in Buffon's *Histoire Naturelle*.

It was through his sister Sarah that Lear gained a practical introduction to the world of illustration, and to the Zoological Society. When he was still only 15, on a visit to Arundel to visit his sister he met Mrs Godfrey Wentworth, whose father was an amateur zoologist and friend of Prideaux John Selby. Shortly afterwards Lear was contributing drawings (one was of the now-extinct Great Auk, which Lear later copied with great success for Gould but was inaccurately attributed to 'J & E Gould') for two volumes of Selby's & Jardine's *Illustrations of British Ornithology*. He was also recruited to produce drawings for wood-engravings for the Zoo's own visitors' guidebook entitled *The Gardens and Menagerie of the Zoological Society Delineated*.

Lear was an artist of considerable talent. He had made his way in his early teens 'colouring prints, screens, fans' and 'making morbid desease drawings, for hospitals and certain doctors of physic'. But he benefitted as well from the instruction of the Zoological Society's draughtsman, William Harvey, who had once been a student of Bewick. Harvey's team of illustrators were encouraged to enliven their work in the dramatic, almost irreverent, Bewick style, to rebel against the static poses and solid profiles of previous zoological illustration. The menagerie of the Zoological Society provided the perfect opportunity to observe subjects alive and kicking, and the tiny woodcuts of the Society's guidebook vibrate with squawking, flapping, screeching birds drawn as never before attempted in a British work of its kind.

The impact of the book, however, was severely limited by its size. The woodcuts are only a few inches square, restricting the amount of detail and background that could be depicted in them. Audubon, at the other end of the scale, had shown in his vast, colourful engravings of birds, the enormous scope for ornithological illustration. His work broke all the conventions of size, style, and technique; his imagination lifted textbook material to the level of a flamboyant, romantic art and set a brave new standard for subsequent bird illustrators.

Inspired by the example of Audubon, his working experience under William Harvey, and the living birds at the Zoological Gardens, Lear decided to embark on his own publication. While still working on the second volume of the Society's guidebook, at a meeting of the Zoological Society on 16 June 1830, he applied to undertake a series of drawings of the parrot collection. In 1831, at the age of 19, he began the publication of his *Illustrations of the Family of Psittacidae or Parrots*. He chose for his format the imperial folio size (22 × 16½in/59 × 42cm) considerably smaller than Audubon's double elephant, but which still gave him the chance to draw most of his subjects life-size. Like Audubon, Lear did not plan for accompanying text: this was to be a purely pictorial study of a single species.

Lear was captivated by the extraordinary and extrovert family of parrots. His drawings are lively, original, and amusing. Some of his birds have individual personality. They twist on their perches in positions never before seen in life-size British bird illustrations, with a mischievous glint in the eye, or looking furious or rebellious; sometimes even forlorn.

Lear, however, as he constantly complained, had not had training as an engraver; and, in an effort to find a medium that would serve his publication without drawing excessively on his limited resources, he turned to the novel technique of lithography.

As a means of reproducing plates in illustrated texts, lithography had not been in use for even a decade. The printing method had been developed by Charles Hullmandel, who later printed Lear's lithographs, and, ultimately, most of Gould's 3,000 plates. Lithography had many advantages over the difficult and complicated process of copper and steel engraving, and even over the woodcut. The lithographic stone is made of a slab of almost pure limestone, which easily absorbs grease. The image is drawn on to the stone with a fine wax pencil or crayon, which attracts the oily ink used in the printing process. After the wax image is washed with the ink, and the excess ink wiped off, a sheet of paper is pressed to the stone, which transfers the ink from the wax lines to the paper. The image that appears on the paper is a mirror-image of the original drawing done on the stone.

Compared with the alternative methods, the technique needs little formal training. The lithographer can draw on a regular piece of paper with the same facility that he can with a pencil or brush. It gives him a far freer hand when describing long, curving lines, particularly compared to what is achievable with a stylus on metal.

Most importantly for Lear, lithography cut out the need for, and the expense of, the middlemen or professional draughts-men, who all too often destroyed the subtlety of a drawing by their clinical interpretation. For the first time, the marks made by the artist could be accurately and directly reproduced on to plates. Once the basic outline of the artist's sketches were reproduced, it was up to a team of colourists, some inevitably better than others, to colour the 200 or so copies of each plate. They would work repeatedly from a prototype provided by the artist, and, although their work varied, the process produced, at a time when the popularity of the English watercolourists was at its height, the closest thing to an original watercolour that a printed book could offer.

Only a few lithographic works had been produced before Lear tried the technique. Hullmandel's *Twenty-four views of Italy* was published in 1818 and was followed shortly by one of the first attempts to use lithography for zoological illustration. William Swainson's *Zoological Illustrations*, which contained 182 plates, 70 of them of birds, was published between 1820 and 1823. But Swainson does not seem to have taken advantage of the ease and freedom which the new technique could afford. His birds are drawn in a style very similar to the engraved figures of his previous books, with the same intricate detail and self-conscious avoidance of any steeply curving strokes. The birds are mostly static, and still shown in profile – a single figure on each page with no background. The size of Swainson's plates also failed to make the best use of the lithographic process: they are as small as regular engravings, with the size of the bird in most considerably reduced. It was therefore left to Lear to realise the full potential of lithography, and to revolutionise bird illustration in the process.

When Hullmandel published his treatise *The Art of Drawing on Stone* in 1824, the possibilities of lithography were better advertised (although Hullmandel was careful not to describe the actual printing process, so that artists would have to come to him for that service), and Lear was one of the very first to be attracted to the technique.

Lear set himself up in his lodgings with the necessary lithographic paraphernalia as well as the heavy stones, which he would carry from Gray's Inn Road to Hullmandel's printing shop in Great Marlborough Street as soon as they were completed. As he wrote to a friend:

'Should you come to town, I am sorry that I cannot offer you a home pro tempore – pro trumpery indeed it would be, if I did not make any such offer – for unless you occupied the grate as a seat – I see no probability of your finding any rest consonant with the safety of my Parrots – seeing, that of the six chairs I possess – 5 are at present occupied with lithographic prints: – the whole of my exalted & delightful upper tenement in fact overflows with them, and for the last 12 months I have so moved – thought – looked at, – & existed among Parrots – that should any transmigration take place at my decease I am sure my soul would be very uncomfortable in anything but one of the Psittacidae.'

It was a brave undertaking for a young man of 18 to attempt to publish, illustrate, and distribute an expensive and novel zoological work alone. Not only was the *Psittacidae* the first volume ever to concentrate on a single species, it was, as Lear himself maintained, 'the first complete volume of coloured drawings of birds on so large a scale published in England, as far as I know – unless Audubon's were previously engraved [here].'

Lear must have received considerable encouragement from the Fellows of the Zoological Society, just as Gould had after him; not to mention the contacts necessary to recruit his 175 subscribers. He was able to make use of parrots not only in the Society's collection, but in the vast private menageries of Lord Stanley, who was then President of both the Zoological and Linnean Societies, and who would later become Lear's principal patron; and in the collections of Sir Henry Halford, 'the eel-backed baronet' and physician to George IV and William IV; of Lady Mountcharles; and of Vigors himself, who was a neighbour of Lear's at Chester Terrace. When live specimens were not available, Lear prevailed upon the extensive collections of the famous dealer in skins, Mr Leadbeater of Brewer Street, Golden Square, and, of course, the Society's own preserver, John Gould.

Lear's *Psittacidae* began to be issued in parts containing three or four plates each from 1 November 1830. From the outset, they were critically acclaimed. On the strength of the first two parts Lear was recommended for an Associateship of the Linnean Society on 2 November 1830 by Vigors, Thomas Bell, and Edward T. Bennett. William Swainson in particular congratulated him in an enthusiastic letter, written from his house in St Alban's:

'Sir, I received yesterday, with great pleasure the numbers of your beautiful work. To repeat my recorded opinion of it, as a whole, is unnecessary but there are two plates which more especially deserve the highest praise; they are the New Holland Palaeornis, and the red and yellow macaw. The latter is in my estimation equal to any figure ever painted by Barraband or Audubon, for grace of design, perspective, or anatomical accuracy. I am so particularly pleased with these, that I should feel much gratified by possessing a duplicate copy of each. They will then be framed, as fit companions in my drawing-room to hang by the side of a pair by my friend Audubon. Of course I can easily perceive when you have been obliged to form your ideas from stuffed specimens. Had I an opportunity I should have great pleasure in giving you a few hints on this subject which might not be useless. Let me suggest that when you want 'relief' for a white bird, the leaves should be coloured: the cold, 'drawing' aspect which the plates have would then be removed.'

Encouragement of this kind alone, however, was not enough to keep the ambitious project afloat. The world of publishing was a ferocious one, and for Lear – a gullible young man with little or no business acumen, and with an expensive project on his hands – it was to prove ruinous. Even later in his career when he was drawing master to the Queen, and published a book of nonsense, a travel book, a volume of landscape drawings and numerous natural history illustrations, Lear made a profit of only £100.

To Gould, though, Lear's enterprise had distinct possibilities. Gould was immediately attracted to the art of lithography for the same reasons as Lear – the relative inexpensiveness of production and the lack of training required – but he also saw great potential for Lear's large-size format and colourful plates, which could, Gould thought, if properly promoted and well managed, prove extremely lucrative. Moreover, if the plates were accompanied by a descriptive text, giving information about the nature and habits of each species figured, the production might appeal to a broader and more conventional audience. A letterpress accompanying the illustrations would give the work a scientific legitimacy that Audubon's work as well as Lear's had lacked.

Gould launched immediately into the fray, and his first part of the *Century* appeared less than two months after the first two parts of Lear's *Parrots*. The almost simultaneous appearance of the two lithographed publications detracted from the dramatic importance of Lear's initiative. It was Gould, not Lear, who was given credit for the concept of an illustrated ornithological work of this nature. The *Proceedings of the Royal Society* completely overlooked Lear's beautiful *Parrots* when it decided that, 'The so-called Century of Birds from the Himalaya Mountains' was 'by far the most accurately illustrated work on foreign ornithology that had been issued up to that period.' 'In size', it remarked, 'the *Century* rivalled the folios of Le Vaillant [sic], but far surpassed them in excellence of the plates.'

Gould approached Lear with an appeal to help his wife master the art of lithography, although the direct influence of Lear's style is not apparent in Gould's plates until Lear began to work on them himself. Lear only sought employment under the Gould regime when he had been forced to abandon the *Parrots*, and the Goulds were about to start on their second publication. Lear himself, however, was aware of the debt that Gould's first publication owed to his initiative. The *Psittacidae*, he always claimed, was 'the first book of the kind drawn on stone in England of so large a size, & . . . one which led to all Mr Gould's improvements.' It soon became painfully apparent, however, that of the two publications by Lear and Gould, only one would prove financially viable. While it must have been galling for Lear, Gould took pains to ingratiate himself with the young artist, and Lear all too easily complied. In a letter to Charles Empson, the Newcastle bookseller, dated 1 October 1831, when he was eight parts into the *Parrots*, and Gould, who had overtaken him by this time, was already ten parts into his own publication, Lear complains:

'My reasons for so soon destroying my drawings were these; though I dare say that they don't appear so rational to any one but myself: I was obliged to limit the work – in order to get more subscribers – & to erase the drawings – because the expense is considerable for keeping them on, & I have pretty great difficulty in paying my monthly charges, – for to pay colourer & printer monthly I am obstinately prepossessed – since I had rather be at the bottom of the River Thames – than be one week in debt – be it *never* so small. – For me – who at the age of 14 & a half, was turned out into the world, *literally without a farthing* – & with nought to look to for a living but his own exertions, you may easily suppose this a necessary prejudice – & indeed – the tarry paying of many of my subscribers – renders it but too difficult to procure food – & pay for publishing, at once. – With Mr. Gould all this is very different – he has sufficient to live on, whether his subscribers pay or not, & can well afford the innumerable little expenses of printing – but for poor I – I have just nine and twenty times resolved to give up Parrots & all – & should certainly have done so – had not my good genius with vast reluctance just 9 & 20 times set me a going again.'

In the same long letter Lear considers the relationship developing between himself and Gould:

'I am glad you like Mr. Gould: – he has been always very obliging to me – although I never knew him till lately: Mrs. G. appears an exceedingly pleasant & amiable woman. to – 'is he at all like you?' – I must say – No – very categorically: setting aside personal appearances – he being stout – & good looking, – & I being ensiform, (speaking botanically,) that is – lanky – & considerably ugly, – we are, as far as I can judge – very opposite. – I believe I am anything but candid: in fact – I am naturally suspicious – & exceedingly reserved, the first good quality arises from my having seen plenty of the evil part of the world from my youth up – the second from being but very little used to company or society – for – excepting Mr. Yarrell – (whom Mrs. Hewitson & Atkinson know,) – to whom I go to study bones & muscles – I don't know a single person in all London to visit intimately.'

Eventually Lear could no longer sustain the psychological and financial burden of continuing his publication alone. By the time he had completed his forty-second plate he was forced to face up to his inadequacies as a business man. Lear had managed to attract 175 subscribers for the *Parrots* but had not the temperament or the means to press them for payment. Gould, in marked contrast, had managed to enlist 298 subscribers for his *Himalayan Birds*, all of whom were successfully pursued by the efficient and persistent Mr Prince. With a great deal of relief, but surely not a little regret, Lear surrendered the copyright of his magnificent lithographs to the hands of John Gould. As Lear told Sir William Jardine on 23 January 1834:

'Respecting my Parrots, there is much to say: – no more numbers will be published to me – the 12th – which you have, being the last. Their publication was a speculation which – so far as it made me known & procured me employment in Zoological drawing – answered my expectations – but in matters of money occasioned me considerable loss. I originally intended to have figured all the Psittacidae – but I stopped in time; neither will there be – (from me) any letterpress.

Concerning the request you make that I would allow these being copied – I have no power either to refuse to comply – since I have sold all right in the volume to Mr. Gould. He purchased of me – the copies left on my hands – and he alone is to be applied to on the point you wish answered.'

Gould's plans to finish the work himself, in which Lear

originally had intended to include every known member of the parrot genus, never materialised. Whether Lear was disappointed is unknown, but certainly Gould realised that he could profit more from starting a completely new project under his own name than by finishing someone else's. Gould was satisfied with simply selling off the remaining numbers he had bought off Lear.

Meanwhile, as the *Century* was nearing completion, Gould was at pains to predict the most popular subject possible for his next publication. In a letter of 17 April 1832 he wrote to Jardine:

'You will perceive by the accompanying Prospectus that I have commenced another work of much greater magnitude [than the Century]; for my own part I should have been more anxious to have gone on with unfigured foreign birds and by that means have added so much the more interest to the science of ornithology, but the greater number of the subscribers to my other work not paying attention to birds generally but limiting themselves to those of our own country, they have frequently reiterated their request that I should commence a similar work on the Birds of ['this country' crossed out] Europe and this has been the only motive for my undertaking so laborious a task.'

There was evidently some dilemma in Gould's mind as to whether a work on the birds of Europe or just the birds of Britain would prove most lucrative. He decided in favour of the former, but was careful to satisfy nationalistic sentiment in his prospectus by confirming that 15 of the 20 plates in each part 'will be devoted to the representation of British Birds, and five to those of the European continent.' Robert Havell Jnr, the London engraver as usual, reported this development to his client Audubon:

'Gould has just issued a prospectus and as soon as I have one I will send it to you, announcing a work on English birds . . . his conceit leads him beyond common sense.'

Although there had been countless works already produced on the European theme, it was a subject that Gould realised would be forever popular among the amateur naturalists and bird-fanciers of Britain. A work on the birds of Europe might be of limited scientific value, as Gould himself confessed – it could hardly uncover any startling revelations – but it would be bound to attract the crucial number of subscribers. Audubon, however, was outraged: 'Here there are at present three Works publishing on the Birds of Europe', he wrote to his friend Rev. Bachman, from London in 1835, '– one by Mr Gould and the others by no one knows who – at least I do not know – Works on the Birds of *all the World* are innumerable – Cheap as dirt and more dirty than dirt . . .'

But it was precisely Gould's unshakeable commonsense, which Havell had so reviled, that was beginning to antagonise the rival American camp, and it was with exasperation that they watched Gould set off for Europe on a reconnoitring trip for his second publication. Gould took the opportunity of taking Lear with him on this ornithological grand tour, and at some point managed to enlist him to work on the Birds of Europe with Mrs Gould. It was an agreement that Lear was to regret bitterly in later years.

In 1832 work on the *Birds of Europe* began. Over the next five years, until its completion, Lear produced 62 of the total 448 plates. In volume Lear's contribution may not have been prolific, but its impact was revelatory. Lear's participation transformed the work of Mrs Gould, which in the *Himalayan Birds* was little more than a continuation of eighteenth-century productions, into dynamic and expressive works of art. Like an ornithological Michelangelo he propelled her limited sense of perspective into the third dimension. He encouraged movement, vigour, and a sense of character in her birds; he instilled an idea of composition in which the subject related to its background, instead of perching in mid-air like a cardboard cut-out. He introduced a sense of subtlety and freedom into her drawings where previously she had only mimicked the technique applied to etching or engraving. There is no doubt that Edward Lear was the first person to understand the art of lithography, and to use it to its fullest potential. It was a legacy that granted the fabled works of Gould their success, and took them into the forefront of nineteenth-century illustration.

Despite these dramatic improvements, not surprisingly Mrs Gould's work falls well short of the plates by Lear. His birds are immediately recognisable, full of character and confidence. Mrs Gould's birds are pictured in less adventurous and often more awkward postures, sometimes regressing to the anatomical inaccuracies and stiffness of her previous work. Certainly, Lear chose the most charismatic subjects for his plates – the owls, eagles, cranes, pelicans, geese, swans and flamingos that became familiar in his nonsense rhymes – and left the less-dramatic little perching birds, such as the thrushes, robins, and wrens to Mrs Gould. Lear's birds fill the page, stalking warily or glowering protectively over a half-dead prey; his eagle owl, perhaps the most famous of all, glares menacingly from the shadows of twilight, its great yellow eyes mirrored by a full moon. Never before, or perhaps since, have birds been portrayed with such vitality, dexterity, and accuracy.

The versatility of Lear's lithographic technique gave the art of bird-drawing a new delicacy of form and texture. Lear showed the sheen of strong flight-feathers, as well as the soft down of the breast; he made the feathers around the neck look ruffled by the wind, and characterised the unique texture of feathers on the feet. Where the technique of etching fell woefully short, he discovered, the art of lithography excelled.

Despite the beauty of Lear's plates, the young artist felt his work for Gould a great strain. He had begun in 1833 to work also on 10 plates for Gould's *Monograph of Toucans*, a single volume of 34 plates on a single species similar to Lear's own original *Parrot* publication, and which was to finish in the same year, 1835, as the *Birds of Europe*. Lear was now working mostly from the menagerie at Knowsley Hall, where he was enjoying the congenial nature of Lord Derby's patronage, and entertaining the Lord's grandchildren. He was also contributing drawings for a number of other zoological works, including Selby's and Jardine's *Illustrations of British Ornithology* and Jardine's *Naturalist's Library*. The pressure from London, however, was unrelenting: 'Gould has been so clamorous lately', Lear wrote to Jardine, 'at my not having done any *Birds of Europe* for him, that I must do a batch for him without further delay.' In a letter to Gould a few months later, Lear, with characteristic self-pity, excuses his poor output:

Knowsley Hall
31st Oct. 1836
hail – snow – frost – & desolation.

Dear Gould,

I got your letter last evening, as I was about to write to you: thank you for it: – I wanted to know how you were going on. As for your replying I never expect that, as I know how busy you are always . . .

To your numerous enquiries, I must answer rather briefly, for I have so bad a cold as to be half blind: add to which so bad a cough & sore throat, that if they dont go I shall be in town much sooner than I expected . . . my eyes are so sadly worse, that no bird under an Ostrich shall I soon be able to do.

I am staying to finish what I had promised Lord Derby I would do before I went away, & shall, I imagine, be home by the end of November, when I shall be most happy to set about anything that you please . . .

Lear was beginning to find the strain of lithography – 'this lampblack & grease work' as he later called it – intolerable, and had made up his mind to become a landscape painter. His worsening eyesight and declining health certainly heralded this decision, but the constraints of working for someone as particular and demanding as Gould could not have helped. 'Lord! Lord! How cross you used to be if a goose's feathers weren't drawn straight!', Lear later wrote to his former employer from the safety of Rome. In 1863, in a note written on a rare visit to England from his self-imposed exile in warmer climes, Lear apologises to Gould for not visiting him because 'all my daylight hours have gone in the service of the old Enemy – Lithography.'

If Lear was unhappy working on the *Birds of Europe*, Gould is unlikely to have been aware of it, or paid him much notice. Time was too precious to allow Gould a thought for artistic temperament. With unswerving determination, he kept his sights on the completion of his publication to the exclusion of all else. 'Mr Gould pursued his onerous task with an intensity of purpose which resulted in complete success', recorded *The Times* about the completion of *Birds of Europe* in August 1837. 'Its reception', continued the article, 'equalled his most sanguine hopes, and its spirited and able author had the happiness to find his labours not unrewarded.' In short, it was a sell-out.

The *Westminster Review* found the evolution of Gould's work since his last publication quite remarkable, and Charles Wood, in his *Ornithological Guide* considered 'the plates in fidelity of delineation and beauty of execution are unrivalled'. The *Athenaeum* saw the *Birds of Europe* as Gould's first serious ornithological work, while the *Analyst* revelled in discovering a work fit to rival the foreign competition. Finally, the eminent ornithologist Charles Lucien Bonaparte agreed that Audubon's supremacy had at last been eclipsed. 'The merit of Mr. Audubon's work yields only to the size of his book,' he said, 'while Mr. Gould's work on the Birds of Europe, inferior in size to that of Mr. Audubon's, is the most beautiful work on Ornithology that has ever appeared in this or any other country.'

Typically, however, there is hardly a word in all the reviews on the labours of Mrs Gould, or on the revolutionary influence of Edward Lear. Only those who had any idea of the inside story recognised the scale of Mrs Gould's contribution. Audubon, who clearly visited Gould sometime in April 1835 wrote to Bachman:

'I have agreed to exchange a copy of my Work with Mr. Gould, for his publications – and have by me 13 Nos of his Birds of Europe etc. I have also purchased a monograph of Parrots from a Mr Lear – When you and I are old men – how pleasing it will be to us to look at these together to quiz them all, and pass our *Veto* upon them!

Mr. Gould is a man of great industry . . . his *Wife* makes his drawings on stone – she is a plain fine woman, and although these works are not quite up to Nature, both deserve great credit.'

There is no mention of Lear's work on *Birds of Europe*; everywhere, in the press in particular, a veil seems to have been cast over Lear's contribution. Worse, two of Lear's plates, of the ptarmigan and the peregrine falcon, are cited as proving John Gould to be 'the first delineator of ornithology of the day'. Lear was subject to even more callous misacknowledgements: in what could generously be described as an oversight, four of the plates in *Birds of Europe*, are referred to at the bottom of the page as being 'drawn from Nature and on Stone by J & E Gould', while the style of the bird and the signature on the drawing itself are patently Lear's. And, where Lear's plates are accurately acknowledged, they are awarded only a cursory 'E. Lear del. et lith.' In addition, some of Lear's lithographs were copied for the *Penny Magazine* where they illustrate articles referred to 'Gould's birds', but do not mention Lear.

Lear drew more than the short straw on the *Birds of Europe*. At best Gould was ungenerous in ensuring that Lear did not receive the praise he deserved; at worst he was deliberately playing on Lear's weakness, and jealously guarding the lime-light for his most ambitious publication to date. Sadly, Lear's contribution to both the *Birds of Europe* and the *Toucans* was forgotten during his lifetime.

It seems that Gould was determined to belittle his talented, but temporary, artist. Prince later wrote to Gould when he was in Australia, complaining that Alfred Newton had been 'far, far too complimentary' about Lear's part in the publication, 'particularly when we know that most of the subscribers are of the opinion that his plates are almost the only exceptionable part of your work.' How Prince could have known the thoughts of almost three hundred people is difficult to imagine, but it was not an opinion held by John Edward Gray of the British Museum, nor of Lord Derby. 'I have just received your Portfolio of MS drawings quite safely,' wrote the former to the latter in January 1844; 'I quite agree with you that they are not to be compared in finish with Lear's, though generally accurate, but then Lear was a Man both for manners in Society & Skill in his Profession not to be easily found or replaced.' Joseph Wolf, the eminent artist who worked for Gould about 20 years later, had his own thoughts on the matter. In his opinion, scientists rarely could distinguish between good, mediocre, or bad art. 'There have been very few among all my acquaintances among naturalists,' he said, 'who could appreciate a drawing if it were ever so well done; and sometimes the better it was done, the less they liked it . . .'

When, eventually, Lear was forced through ill health and exhaustion to exchange England and 'the old Enemy, lithography' for a life of watercolour painting in Rome, he was desperate to keep in touch with old acquaintances. He seemed to be haunted by a fear of being forgotten by his friends and Gould became one of the many objects of an obsessive correspondence.

Lear's letters to Gould in the years after they parted company were long, funny, and warm; but they were also the letters of a nostalgic and lonely expatriate. Time and again Lear tried to tempt Gould to visit him in Rome; always he pleaded for a letter: 'I am anxious to hear from you . . .'; 'I beg you will write to me after you get this'; 'Pray send a line . . . as I should like to know how you are going on, – as I don't forget I have owed sundry dinners to your lithographic birds.'

None of Gould's answers has survived, but they clearly fell well short of Lear's hopes. 'I was always a bad correspondent,' berated Lear, 'but surely you are still more unconscientious, for when I do write, you answer me by a short scrawl – only one word of which out of every 2d can I decipher, & I have kept your last and only epistle to see if I can't sell it as an ancient hieroglyphic.'

Gould, however, was the last person Lear should have expected to engage in a chatty, long correspondence. Virtually all Gould's letters, even those to his children, are brief and matter-of-fact. He had little time for flippancy or gossip, let alone the nostalgic ramblings of an ex-employee. It is difficult to imagine the sense of kinship between the two men which Lear implies in his letters to Gould. There could scarcely have been two people more different from each other in character: the one motivated by a keen sense of order, rationality, and discipline; the other driven by his emotions and whimsical fancy. Where Gould was shrewd, cautious, even cunning, Lear was gullible, naive, and often incompetent. What Lear revered in Gould – his remarkable business sense and financial efficiency – Gould probably found exasperatingly lacking in Lear.

In the beginning, one feels, Lear must have been relieved and excited to find someone with as equally compelling an interest in birds as himself. But there was no sentimentality in Gould's attitude to birds, and he was, for the times, extraordinarily misanthropomorphic in his descriptions of them. His was the discipline and the drive of the scientist; Lear was first and foremost an artist. Mrs Gould may have been able to bridge the gap with a foot in both camps, but the two men must inevitably have begun to polarise themselves from one another.

Gould's self-confidence and his clarity of purpose were anathema to Lear, whose life was buffeted by self-doubt and a scepticism verging on paranoia. Lear was volatile and moody – and one can imagine the difficulty Gould faced if he criticised a drawing – but he was capable of great emotional generosity and open-heartedness. Gould, on the other hand, was reticent, rarely emotional, and often insensitive. Lear must have found this side of Gould the most frustrating and hurtful; he could not understand someone who did not think people were important. Time and again Lear's long, witty, and entrancing letters were responded to, if at all, by what Lear called 'a nasty abortion of a note'. But Gould did not have time to fuel his friendships, unless they were of some direct advantage to his work. And Lear took his own inability to captivate Gould as a personal failure. Years of being dismissed by Gould while Lear was lonely and homesick in Rome eventually took their toll, and finally he turned against one of the friends he had been trying most persuasively to woo.

The famous trip to Europe, which Lear had constantly referred to in his letters as if it were an experience which united him with the great ornithologist, became the bitter disappointment of a friendship *manqué*. Lear was doubtless unhappy during much of the time he worked for Gould; he was also constantly unwell, often in desperate financial straits, and had been finding lithography an increasing strain on his eyesight and his nerves. Gould could not be blamed for all of this. Lear's bitterness stems as much from the consequences of his own self-deception in regard to his relationship with Gould, as

Gould's insensitivity. Isolated in a foreign country, and longing for all that had once been familiar, if not at the time congenial, to him, it was easy for Lear to expect of Gould what he could never give. Finally, Lear resigned himself to Gould's impenetrable isolation, and, in a letter written in 1863, came as close as he could to an objective description of his former employer and erstwhile friend: 'A more singularly offensive mannered man than G. hardly can be: but the queer fellow means well, tho's more of an Egotist than can be described.' It was an opinion that not a few of Gould's associates would have shared.

CHAPTER 5

A Dram-Bottle from Darwin

B EFORE THE END of his mammoth production *Birds of Europe* and his *Monograph of Toucans* ever came into view, Gould began planning his next major project. He had been receiving an extraordinary number of new birds from Elizabeth's brothers, Stephen and Charles Coxen, who had emigrated to Australia some years before. It was clear from their reports that much of this strange and exciting land had barely been explored, let alone ravaged by the ubiquitous ornithological collector. As Gould rightly assumed, vast areas of *terra incognita* held enormous possibilities for new and exotic species. Australia beckoned to him like an uncut diamond. On 2 November 1836, a year before the last part of *Birds* was published, Gould wrote to Jardine of his latest ambition: 'Would not a work on the Birds of Australia be interesting? I have a great number of new and interesting species to make known etc. but have an idea of making it my next illustrative work. I have even some serious intention of visiting the colony for two years for the purpose of making observations etc. I have several plates of my new species already drawn and coloured, and will some day take an opportunity of letting you see them.'

Expeditions to the far-flung corners of the world in search of scientific novelties were in vogue that season as they had never been before. Exactly a month earlier, a ship had docked at Falmouth after a five-year voyage surveying for His Majesty's Government. It had successfully conducted its official task of charting the South American coast, of assessing potential harbours, transport systems, and inland waterways around the world. It had also returned loaded with magnificent botanical, zoological, and geological collections for the Crown.

One man on the voyage had drawn particular publicity during his absence abroad. He had been taken on in an unofficial capacity as Captain's companion, but, like everyone else on board, from ship's surgeon to midshipman, he had made the best of this great opportunity by gathering remarkable collections of insects, plants, birds, and fossils. Barrels containing the dried, wrapped, skinned, and pickled specimens of this private collection were dispatched regularly back home to John Henslow, Professor of Botany at St John's College, Cambridge. Long before the expedition returned, it was Henslow who began to spread the word that these collections were indeed something to write home about. By the time *HMS Beagle* reached her final port, Charles Darwin had been elevated from obscurity to the status of reputed naturalist.

As yet, though, there was no indication that this particular voyage of the *Beagle* was to be considered one of the most important events of all time. It encouraged other energetic young scientists to engage in similar expeditions, and it showed armchair naturalists that the opportunities available to the man in the field far outweighed the advantages of staying at home. But for the time being, the impact of the voyage was limited to the kudos of returning laden with booty and important information about new trade routes.

John Gould was no doubt spurred on by the success of Darwin, Captain FitzRoy, and the *Beagle*'s other collectors, and undertook an expedition of his own to Australia. He knew that firsthand experience and observations could prove invaluable in collating, describing, and illustrating a definitive work on Australian birds. He, more than any other ornithologist, knew what to look for and would be able, with the help of a few hand-picked collectors, to assemble a substantial collection in a fraction of the time it would take should he stay home and await specimens sent by less reliable amateurs.

What Gould had never fully recognised, until the rapturous reception of the *Beagle*'s return, was the international acclaim attached to this kind of voyage, which could, it seemed, make a hero out of anyone. Naturalists of any repute were soon expected to go on expeditions, and those who did not were criticised for their lack of field experience. For Gould, eager to shake off the connotations of his previous post as a dusty museum curator, and besieged by allegations of being an 'indoor naturalist', time spent collecting in the southern hemisphere would silence even his most vociferous opponents.

When Darwin delivered his collection of birds and mammals to the Zoological Society for identification three months after his arrival back in England, Gould probably welcomed the opportunity of finding out more about the voyage, particularly the month or so Darwin spent in Australia and Tasmania. Gould would also have angled for contacts who might help him with his prospective expedition. It could have been Darwin himself who introduced Dr Benjamin Bynoe, the *Beagle*'s surgeon, to Gould. During the next voyage of the *Beagle*, Bynoe contributed a number of new finds for the ornithologist's work on the *Birds of Australia*.

Gould probably never imagined that his meeting with Darwin would be one of the most significant events in scientific history; that what he was to uncover in Darwin's bag of birds was to spark off the theory of evolution. 'I must not omit to tell you,' wrote Gould nonchalantly in a letter to Jardine on 16 January 1837, 'that Mr Darwin's Collection of Birds (made during the late survey under Capn FitzRoy) are exceedingly fine; they are placed in my hands to describe; some of the forms are very singular particularly those from the Gallipagos [sic]. I have one family of ground Finches in which there are 12 or 14 species all new.'

For it was Gould who showed Darwin his first evidence of the possible transmutation of species. The very specimens to which Gould refers so casually in his letter, particularly those which came to be known as 'Darwin's finches', were to spark off one of the most controversial theories of all time. It was Gould, not Darwin, who identified them. Contrary to popular myth Darwin was not thunderstruck by the theory of evolution during his voyage on the *Beagle*. The revelation came, if it can be accorded to any single event, when he was back in England, reviewing his collections with one of the best ornithologists in the country.

As a naturalist, the young Charles Darwin was far from fully qualified. He was expecting to become a country parson on his return to Britain, and was only assigned to the *Beagle* expedition through the recommendations of Professor Henslow, his mentor. The position of Captain's companion had been offered to at least four people before Henslow suggested Darwin as a possibility. Darwin, as his mentor suggested, made the most of his time abroad, compiling copious notes about the geography, geology and wildlife he saw. The heretical ideas that were to label Darwin as 'the man who killed the Creator' could not have been further from his thoughts. Midway through the voyage he wrote to a friend, 'I often conjecture what will become of me; my wishes would certainly make me a clergyman.'

Darwin, by the time he reached the Galapagos Islands, was not in a particularly receptive or philosophical frame of mind. The *Beagle* was nearing its fifth and final year and Darwin was tired, homesick and chronically sea-sick. The last leg of the expedition was proving more than a trial for the young naturalist. The entries in his diary for Tahiti, New Zealand,

and South Africa are little more than dutiful records, lacking enthusiasm and inspiration.

The *Beagle* stayed only five weeks in the Galapagos. Isolated in the middle of the Pacific Ocean, 3,200 miles from Tahiti, and with a sea-crossing of almost a month stretching out before him to the next port of call, Darwin's stay in this inhospitable volcanic outcrop must have been one of the lowest points for him on the entire trip. That he managed to summon enough energy to make any collections at all during his stay is something of a miracle; had he been more punctilious in his methods he would have saved history, as well as himself, a lot of time and trouble when it came to corroborating the theories that were to bring him such fame, and infamy, many years later.

Captain FitzRoy was also getting on Darwin's nerves by this time. Four years at the Captain's table were taking their toll. The two men, the former a steadfast Tory, the latter a dedicated Whig, had crossed swords on several occasions. The Galapagos gave rise to one particular dispute between the two men – one which was to prove, much later – to be at the core of the problem of evolution. FitzRoy had been struck by the peculiar appearance of a group of brownish-black birds living in the islands that resembled finches. He noticed that some had beaks that seemed to be perfectly created for the needs of the locality. The cactus finch, for example, has a long beak designed to feed on the flowers of the prickly-pear cactus; while its remarkable mangrove cousin uses its beak to manipulate twigs to extract insects from deep within the bark of trees. FitzRoy immediately guessed that what he saw was a new species. He assumed, however, that rather than evolving over the ages to suit their environment, they had been 'ready-made' by God to the perfect design needed for where He placed them.

Darwin was not quite so orthodox in his interpretation. From his study of geology he knew that the world was much older than Archbishop Usher's famous calculation that there were a mere 4,004 years between creation and the birth of

Christ. He could see, moreover, that the Galapagos were much younger than the continent around which he had been travelling. The natural inhabitants of the Galapagos then, must have somehow strayed across from South America, and colonised the archipelago by chance. And their strange adaptations, such as the finches' beaks, must be a response to the new and strange environment in which they found themselves.

There was nothing too outrageous in this conclusion. Adaptation as a concept was perfectly acceptable to the nineteenth-century naturalist – they had, after all, had plenty of opportunity to observe the changes that an English climate wrought on the coats and plumage, as well as the size, of live birds and animals brought from hot countries. And the cross-breeding of domestic animals and livestock had long since proved that such changes could be artificially induced. But adaptation was considered only ever to be superficial. It might result in a variety of a certain species, but it could never result in a completely new species. Man and the environment might be able to create a variety, but only God could create a species. Accordingly, all the species on earth had changed only superficially, if at all, since they were placed there by God in the beginning. Darwin's conclusion then, in direct contrast to that of his even more orthodox Captain, was that the Galapagos finches could only be varieties, and were therefore not worth collecting carefully.

There were other extraordinary, dramatic examples of transmutation that Darwin encountered on his voyage and overlooked, quite apart from those found in the Galapagos. One bird that was shot in Patagonia was almost fully digested by the hungry crew of the *Beagle* before Darwin realised it might hold some scientific significance and should be preserved for identification back in London. It, too, eventually became a vital piece in the evolutionary jigsaw that Darwin was to assemble with the help of John Gould at the Zoological

Society. Darwin had heard reports from the native Gouchos in Patagonia of a very rare ostrich-like bird, which they called 'Avestruz Petise'. He tells the bizarre story of its discovery in his account of *The Voyage of the Beagle* (1839).

'When at Port Desire, in Patagonia (lat 48°), Mr Martens shot an ostrich; and I looked at it, forgetting at the moment, in the most unaccountable manner, the whole subject of the Petise, and thought it was a two-third grown one of the common sort. The bird was cooked and eaten before my memory returned. Fortunately the head, neck, legs, wings, many of the larger feathers, and a large part of the skin, had been preserved. From these a very nearly perfect specimen has been put together, and is now exhibited in the museum of the Zoological Society. Mr Gould, who in describing this new species did me the honour of calling it after my name, states, that besides the smaller size and different colour of the plumage, the beak is of considerably less proportional dimensions than in the common Rhea; that the tarsi are covered with differently-shaped scales, and that they are feathered six inches beneath the knee. In this respect, and in the broader feathers of the wing, this bird perhaps shows more affinity to the gallinaceous family than any other of the Struthionidae.'

It is impossible to imagine such an accident ever happening to the meticulous John Gould, but Darwin was no ornithologist, and, as a naturalist, he had had little training in the disciplines of collecting or identification. He was an enthusiastic amateur playing a professional's part. Like the little half-eaten ostrich so expertly reconstructed by Gould back in England, much of what Darwin had seen on the voyage had to be reassessed and re-evaluated after the event. Only as Darwin began to form his theories did he realise the error of his ways on board the *Beagle*.

Much of the evidence that he later used to support his arguments had to be provided from collections other than his own, including that of his servant at the time.

There is only one suggestion in his *Ornithological Notes*, written on board the *Beagle* after his visit to the Galapagos, that Darwin was significantly puzzled by the species he found there. While he was cataloguing his mocking birds, Darwin remarks:

'I have specimens from four of the larger islands. . . . The specimens from Chatham and Albermale [sic] Isd. appear to be the same; but the other two are different. In each Islad. each kind is *exclusively* found: habits of all are distinguishable . . . When I see these Islands in sight of each other, & possessed of but a scanty stock of animals, tenanted by these birds, but slightly differing in structure & filling the same place in Nature, I must suspect they are only varieties. The only fact of a similar kind of which I am aware, is the constant asserted difference – between the wolf-like Fox of East & West Falkland Islds. – If there is the slightest foundations for these remarks the zoology of Archipelagos – will be well worth examining; for such facts [would – (inserted)] undermine the stability of species.'

But it by no means establishes that he was beginning to question the stability of species. Quite possibly he was simply trying to prove it right. In any case, we hear no more of Darwin's tentative thoughts on the subject for the rest of the journey. Like the finches, the Galapagos mocking birds, Darwin concluded, must only be varieties.

When Charles Darwin delivered his boxes to the Zoological Society on 4 January 1837, he could barely have imagined the excitement they would generate among the Fellows. The Society's naturalists immediately began work on the identification of the unusual collection, and only six days later, at the next regular scientific meeting, they were able to announce a number of exciting discoveries. The *Morning Herald*, as well as two other newspapers, and the *Athenaeum*, carried the story:

'ZOOLOGICAL SOCIETY – The ordinary meeting was held on Tuesday evening; W. B. Scott, Esq, in the Chair. On the table was part of an extensive collection of mammalia and birds, brought over by Mr Darwin, who accompanied the Beagle in its late surveying expedition in the capacity of Naturalist, and at his own expense, a free passage only being allowed by the Government. Of the former there were 80, and of the latter 450 specimens, including about 150 species, many of which are new to the European collections . . . Several species of the mammalia were explained by Mr. Reid, amongst which was a new variety of Felis [cat], named F. Darwinnia, with several opossums. Mr. Gould likewise described 11 species of the bird brought by Mr Darwin from the Gallapagos [sic] Islands, all of which were new forms, none being previously known in this country.'

These 11 species of birds identified by Gould were none other than the Galapagos finches. Ironically, the observations of Captain FitzRoy, the orthodox Creationist, had been proved right. It was a remarkable discovery on Gould's part, given that, apart from their extraordinary beaks, many of the finches simply did not resemble finches. The little birds had assumed the role of birds that were absent from the area. They had fulfilled various ecological niches which were occupied on the American continents by other birds. Small wonder then, that the inexperienced eye of Charles Darwin mistook the warbler finch for a 'wren' or a warbler, and the cactus finch for a member of the oriole and black-bird family. In fact, Darwin initially distinguished only six of Gould's eventual 13 separate forms as members of the finch family.

As yet, however, there was no indication from the accounts

available to Gould that these different species of finch depended for their characteristics on their different localities. The labels Darwin had attached to them did not say where the specimens were found; he simply had not had any reason at the time to think that geography was important.

It was only when Gould declared on 28 February 1837, that Darwin's island specimens of the Galapagos mocking bird differed so profoundly that they had to be categorised as three completely distinct species, that the alarm bells began to ring. Fortunately, because Darwin had noted where the mocking birds could be found, it was possible to show that each new species was confined to its own island. Shortly after Gould's publication of this information, Darwin moved to London to be closer to what was proving to be a hotbed of scientific activity. Darwin was beginning to realise that the naturalists at the Zoological Society might need more information than he had provided in his notes.

Previously Gould had been making other extraordinary discoveries from Darwin's collection. On 24 January he identified six new species of birds of prey – two of which came from the Galapagos. On 14 February he named two new species of goatsucker, a new species of kingfisher, and two new swallows, one of which (*Hirundo concolor*), came from the Galapagos. Ultimately, Gould was to name 25 of Darwin's 26 Galapagos land birds as new and endemic; and three of the 11 water birds, usually expected to have a far wider geographical distribution, as belonging exclusively to the Galapagos. Clearly this bare volcanic archipelago was proving to be, as Darwin later described it, 'a little world within itself.'

It was not until after Darwin's meeting with Gould at the Zoological Society to discuss the ornithological collection in early March, however, that the penny finally dropped. Among Darwin's papers at Cambridge University Library there is an extraordinary sheet of paper that bears testimony to this historical visit. It is covered with a hasty scrawl in Darwin's hand. On both sides is the word 'Galapagos', scribbled in pencil vertically over the text. The piece of paper records a list of all the new species identified by Gould from the Galapagos, with, alongside them, notes about their alliance with species on the American continent. Although Darwin had guessed that the Galapagos birds originated from there, because the Galapagos were so young, it was not something of which he could be certain. Gould however, with his intimate knowledge of South and North American bird species, immediately saw the connection.

The order of the list of land birds is almost exactly the same as appears in Darwin's 1839 *Journal of the Voyage of the Beagle*, so these notes must have been used by Darwin as the basis for his subsequent publication. From the very first entry at the top of the list it is clear that this discussion was to be for Darwin revelatory. The first word – 'Buteo?' – refers to Gould's classification of Darwin's Galapagos hawk. Darwin, from his observations of the bird's feeding habits and its tameness alone had assumed it belonged to the *Caracaras* or *Polyborus* family. Unaware of these observations however, Gould had identified it, physiologically, as a member of the *Buteos*. Both were, in a sense, right, although Darwin initially was sceptical. He wrote at the top of his list 'Buteo' with a question mark. Gould, though, having had from Darwin's eyewitness account of the bird's behaviour, subsequently agreed that the Galapagos hawk shared characteristics of *both* the *Polybori* and the *Buteos*. It was an observation which was to emerge again among the germinating ideas and jottings of Darwin's first notebook on *The Transmutation of Species*. Finally, in the second edition of his *Journal of Researches*, the observation emerged fully fledged as testimony to Darwin's theory of convergent behavioural evolution.

On the other side of this extraordinary piece of paper, written sideways at the bottom of the page, is a list of Gould's names for Darwin's finches. Three notes, written horizontally have been added. The first remarks, '26 true land birds, all new except one: rallus'. Underneath it a second note reads, 'Capt FitzRoy Parrot beaked finch comes from James Island'. The third is written in pencil and partially erased. It reads, 'No specimens [or species] from James Island'. This is the very first time that Darwin associates geography with biology. The two notes referring to James Island suggest that he has just started to identify different species by their specific island localities. Gould and Darwin must have discussed not only the alliance between the species in the Galapagos and the species on the continents, but the reliance of the Galapagos species for their individual characteristics on their particular environment. The theory of the adaptation of species had just taken seed.

This single sheet of paper, though, tells us even more about the dramatic meeting between Gould and Darwin. The sheet is of narrow-ruled paper and bears the watermark 'Reeve Mill 1836'. It was the kind of notepaper being used at the Zoological Society at the time, and completely unlike anything ever used by Darwin. On one side, underneath the writing in ink, is a small pencil drawing of an animal. Apparently, Darwin, calling at the Society on what he expected to be a routine visit, was not equipped with paper. Stimulated by Gould's remarks, he reached for a piece of paper lying near to hand – presumably one that was being drawn on by one of the zoological draughts-men working at the Society at the time – and proceeded to write on the reverse side of the drawing. When he had filled that side, he flipped the page over and continued to make notes. Scribbling straight over the pencil sketch and, in his haste, spelling phonetically, most noticeably dropping the silent 'p' in 'psittacula', the twelfth of his finch species to be commented on by Gould. Later Darwin was to refer to this in his pocket journal as the moment of his conversion.

A few days after this impromptu but cataclysmic dialogue, Gould and Darwin presented papers together at the Zoological Society meeting of 14 March on the two species of ostrich

Darwin had found in Avestruz in Patagonia. One was the common ostrich; the other was the smaller 'petise' version that Darwin had snatched from the jaws of his shipmates. Gould identified them as distinct and separate species, the latter being new to science. The following day, Darwin made the first evolutionary entry in his current pocket notebook (known as the 'Red Notebook'): 'When we see Avestruz two species. certainly different, not insensible change. – yet one is urged to look to common parent? why should two of the most closely allied species occur in the same country?'

One cannot overestimate the importance of Gould's identifications in establishing foundations for Darwin's subsequent theories, as well as the part they played initially as catalyst. For there was much speculation and controversy among professional taxonomists at the time about the distinction between a variety and a species. But Gould succeeded in persuading Darwin that his were the right decisions; Darwin's entire case rested upon Gould's evidence. If it could be proved that new species had emerged since the creation of the world, then the story of creation would be thrown into doubt. Creatures were not necessarily placed on the earth in the form in which they appeared now. Even *mankind* might have developed from a different species. And if species did not owe their origins to God, was there a Divine Creator at all?

It might be with these frightening possibilities already in mind that Darwin wished to establish the credibility of these new Galapagos species by publishing an illustrated work on them – before he entered upon the exposition of his theories of evolution. Gould saw potential of a different kind in Darwin's collection. Ever keen to have a finger in any promising pie, he urged Darwin to publish as early as January 1837. Darwin, however, waited until April – after his March meeting with Gould – before deciding to go ahead.

Perhaps because the Galapagos was bound to be an explosive issue, and he was still unsure of his findings, Darwin decided to publish a more general *Zoology of the Beagle's voyage on some uniform plan*. Darwin acquired the important stamp of scientific approval for his collection by designating a reputed naturalist to work on each category for publication. Published eventually in five volumes on fossil mammalia, mammalia, birds, fish, and reptiles, respectively, the *Zoology* openly claimed for science some of the many new species that had been named from the *Beagle*'s collection.

Darwin was eager to emphasise the reputation of the scientists who had worked on the descriptions for each volume. With the weight of these great taxonomists behind him, he could convince the rest of the scientific fraternity of the validity of the new species he had discovered. Resolving the species/variation question in this way was the first step towards introducing the world to the theory of evolution. For even he, the most sceptical perhaps of all, had been persuaded in the end by the professional judgement of the men at the Zoological Society, and most particularly by John Gould.

If Gould was excited by these discoveries, we hear nothing of it. Certainly he refused to implicate himself in the development of a theory in which he had played so great a part. And, while Darwin often made use of Gould's observations on the characteristics and distribution of certain birds in his *Origin of Species*, Gould never broadcast his role as one of Darwin's sources. As late as 1864, five years after the appearance of *Origin of Species*, Alfred Newton remarked to his brother Edward on Gould's extraordinary avoidance of the entire question, 'It is most amusing to see how anxious he is to avoid committing himself about Darwin's theory. Of course, he does not care a rap whether it is true or not – but he is dreadfully afraid that by prematurely espousing it he might lose some subscribers, though he acknowledged to me the other day he thought it would be generally accepted before long.'

With his sights firmly set on his own career and the publication of his next work, Gould simply ignored the earthquake rumbling beneath his feet. He continued with his plans for an expedition to Australia while Darwin was in the process of conceiving one of the most important discoveries in scientific history. Gould accepted the commission for illustrating the volume on birds for Darwin's *Zoology* but left before he saw it completed. In pages i and ii of the advertisement for this volume, Darwin explains Gould's part in its production and the untimely departure which left Darwin with so many questions still unanswered:

'When I presented my collection of Birds to the Zoological Society, Mr Gould kindly undertook to furnish me with descriptions of the new species and names of those already known. This he has per-formed, but owing to the hurry, consequent on his departure for Australia, – an expedition from which the science of Ornithology will derive such great advantages, – he was compelled to leave some part of his manuscript so far incomplete, that without the possibility of person communication with him, I was left in doubt on some essential points. Mr. George Robert Gray, the ornithological assistant in the Zoological department of the British Museum, has in the most obliging manner undertaken to obviate this difficulty, by furnishing me with information with respect to some parts of the general arrangement, and likewise on that most intricate subject, – the knowledge of what species have already been described, and the use of proper generic terms. . . . As some of Mr. Gould's descriptions appeared to me brief, I have enlarged them, but have always endeavoured to retain his specific character; so that, by this means, I trust I shall not throw any obscurity on what he considers the essential character in each case; but at the same time, I hope that these additional remarks may render the work more complete.

The accompanying illustrations, which are fifty in

number, were taken from sketches made by Mr. Gould himself, and executed on stone by Mrs. Gould, with that admirable success, which has attended all her works.'

On several occasions, Darwin had to ask Prince to supply him with information that Gould had not given him. 'Will you have the kindness to inform me,' wrote Darwin to Gould's secretary on 21 January 1839, 'what generic or sub-generic name, Mr Gould has given to the Goat-Suckers of the United States. – Mr. Gould has left a blank for this name to be put in, in one of his M.S. pages. – Will you send me an answer by the Bearer, or a line by the 2d post. – I was very glad to see by the Newspapers that Mr. Gould had arrived safely at Hobart Town. Yours truly, Chas. Darwin.'

Joseph Wolf, who later worked as an artist for Gould, once said, 'to have been in any way associated with Charles Darwin is an event in a man's life.' Gould it seems was only too eager for the event to be over. Before leaving for Australia, Gould received a gift from Darwin. In a letter dated Good Friday, 13 April, 1838, he thanked him for it:

'My Dear Sir

I received with much pleasure the token of your friendship which you have sent me this morning in the way of [a] dram Bottle and shall treasure it up until I reach the wilds of the beautiful country I anticipate so much pleasure in exploring. Your Gift will there daily remind me of your friendship and the many pleasing chats we have had in our favourite pursuits.

G Nat History &tc will also be of interest to me as an evidence that our intercourse has been productive of none but the most amicable feelings, that your [two words illegible] [eyes the blessing of friends] [illegible]

Ever yr J. G.'

The tone of Gould's letter was typically formal and cursory, given the long hours and the revelations they had shared. But Gould had quite possibly thanked Darwin for the wrong present. There is no record of a dram-bottle from Darwin ever having been in the Gould family; there is, however, a beautiful silver compass inscribed 'Mr. Gould from C. Darwin Esq'.

It is amazing, with hindsight, to see the awakening of the theory of evolution dismissed by Gould as nothing more than a number of 'pleasing chats'; but Gould was not a revolutionary and never claimed to be; his entire life and all of his works were designed to get him accepted by society and by science, not rejected by them. He would not risk all he had so far gained by advocating a theory as controversial as Darwin's. If he could see the sense behind Darwin's arguments, it is unlikely that his subscribers would; and it was for their tastes that he catered. It was far safer politically and economically to sit on the scientific fence.

Gould got what he considered to be the best out of Darwin and the *Beagle*: he had published a number of scientific papers on the exciting discoveries and new species contained in Darwin's collection, and he had produced another important illustrated work. He would reap the glory without incurring the criticism. His departure for Australia came at just the right time.

CHAPTER 6

Birds of Passage

B Y LEAVING FOR AUSTRALIA, Gould was not only sidestepping the controversy he had helped to agitate with Darwin; he was preserving for himself a niche which would forever distinguish him from the motley and profusive competition of all other ornithological illustrators. England had been swamped in recent years by publications of every sort on birds, and Gould was not the only one to feel the pressure of increasing competition. 'London is just as I left it,' railed Audubon resentfully to his friend, the Reverend Bachman, 'a Vast Artificial area, as well covered with humbug, as are our Pine Lands and old fields with Broom grass. Swainson has just finished his Birds of Europe. . . . Yarrell is publishing the British birds quarto size – and about one thousand other niny tiny Works are in progress to assist in the mass of confusion already scattered over the World.'

Australia, alone, remained unexploited. Virtually nothing so far had been published on its birds. Gould knew, long before he ever decided to set sail, that the province of Australia had fallen miraculously into his lap, and that he would be able to publish from the 'rich ornithological productions' of this country a series of works which were beyond challenge. While others might attempt publications on the birds of Europe, or produce monographs based on species preserved in all the great collections, no one had the ability or the connections to tackle Australia. 'You are perhaps aware', wrote Gould grandly to William Swainson in January 1837, 'that I have two of Mrs Gould's brothers in Australia engaged in collecting the natural product of that fine country, nearly the whole of which are consigned to myself and that consequently I possess perhaps greater facilities than most persons for the production of a work of this description. I think I shall add at least as many more species as are at present known.' His boast did not succeed in warning Swainson off what Gould now considered to be his territory. On 2 May 1839 Prince wrote to Gould in Australia warning him of Swainson's impending visit: 'Mr. Yarrell has just called and requested me to tell you that Mr Swainson is going to Australia whether to collect or to settle is not known at present but he considered it of importance that you should know it as quickly as possible not that either he or I think it can in any way affect you as you have so good a start and his wonted and well known irregularity of publication will surely militate against him.'

On the strength of the parcels sent back to him by his brothers-in-law, and what Gould described as his 'own exceedingly rich collection, perhaps the finest extant', he had already begun to publish a volume entitled *A Synopsis of the Birds of Australia and the Adjacent Islands*. This was a small, imperial octavo-size volume containing eventually 73 plates, mainly of just the heads of birds. It was designed both to whet commercial appetite for the more ambitious 'large work on the Birds of Australia', which he had in mind, and to establish the scientific validity of the enterprise. Certainly it was much more to scientific taste than the huge, glossy volumes Gould had published before. The synopsis was, as Gould bragged to Jardine, 'very much approved of', although Jardine himself still had reservations about the cost of the production and the lavish colouring. To most people it was only the text that could be found lacking. 'The figures of the heads are beautifully done,' noted Selby, '[and] are quite sufficient to identify the species, [but] I wish he had added a little more to the letter press as not a single word is said about habits, a description of the plumage alone being given.'

Gould was well aware of the descriptive shortcomings of his synopsis – there was simply not enough information available on Australian birds – but, cautious as ever, he needed to prove the strength of the market before committing himself to the 'grander work in prospect'. In the event, so encouraging was the reception of his synopsis, so fertile the ground for his proposed work, that Gould was incited to commit an uncharacteristic error of rashness. Despite announcing the projected two-year trip to the southern hemisphere expressly for this purpose, Gould began to publish the first parts of his *Birds of Australia* before he had even left. 'Gould is publishing the Birds of *Australia* from stuffed Skins', wrote Audubon contemptuously to Bachman.

It did not take long for Gould to realise that it would have been better to have waited. The 'strange and anomalous nature' of the 'few extraordinary species' he had before him pickled in spirits or dried as skins confounded the imagination even of an ornithologist as instinctive as Gould. He simply had no idea what these strange creatures could look like in the flesh. Many of the specimens that had found their way back to England only added to the myths circulating in the popular imagination about the fauna of the southern hemisphere. Birds of paradise skins that had been prepared in the traditional way by natives of New Guinea had had their legs removed; for a considerable time even respectable scientists thought there might be a land where legless birds spent their entire life on the wing. The kiwi, on the other hand, which was only discovered in 1813, was thought so ludicrous as to be a certain hoax. With nostrils at the end of a proboscis-like beak, conspicuous ears and unable to fly, it seemed more mammal than bird. The nocturnal ground parrot, or kakapo, also seemed the stuff of legend: a huge, flightless bird – the largest parrot in the world – which scavenged clumsily on the forest floor.

Such creatures defied all Western ornithological experience, and made the work of illustrating them as they were in life impossible. There was no alternative for Gould but to suppress the 20 plates he had published, refund his subscribers, and go straightaway to Australia to find out what they looked like. It had been one of the few mistakes Gould had ever made. 'You will perceive', he noted touchily to Jardine, when Jardine implied that the subscribers had borne the cost of the cancellation, 'that I am a looser [sic] and that to a considerable amount, and not the former subscribers.'

Most of the cancelled parts, of which there were two, were returned, and Gould arranged to give those subscribers the first part of the new series in lieu. Alternatively, as he explained to Lord Derby, 'the cancelled parts, may, if you please, be bound together, and entitled *Illustrations of Birds from Australia*, and from the few copies in the hands of the Public, they will some day be of value, though, it is true, more to the Book-Collector, than the Naturalist.' Gould's unerring commercial nose proved right – today the suppressed parts of *Birds of Australia* are rare indeed and valued at around £30,000. But one cannot deny that a more scientific decision would have been to try to make the return of the offending plates obligatory, so that only the accurate, authentic figures were permitted circulation.

However necessary a trip to Australia began to seem, it was not a voyage to be undertaken lightly. It was only 50 years since the first white settlement had been established at Port Jackson, and, although some semblance of order and 'civilisation' existed in the larger towns like Sydney, many of the outposts were still dangerous places to visit. Melbourne, for example, was a village only five years old. Reports of escaped convicts and savage blacks were rife, although no doubt exaggerated for the benefits of the pink-cheeked 'pomegranates', or 'poms', arriving fresh from England. Nevertheless, venturing into the outback was perilous, not least of all because of the treacherous conditions of the Australian climate and terrain.

Very little of the vast continent had been explored. It was only nine years since Charles Sturt had returned, starving and almost blind, from discovering the Darling and the Murray Rivers; it was to be another five years before he ventured towards the centre of Australia in search of that mythical inland sea. Many men, natural scientists included, had been lost in the wastes. Richard Cunningham, a botanist who had been trained at Kew by William Townsend Aiton, elder brother of John Townsend Aiton (who had trained Gould as a gardener at Windsor Castle) had been murdered by Aborigines only a short time before Gould's departure while collecting on an expedition along the Murray River.

John Hutt, Governor of Western Australia, was thrilled that it was to be an Englishman who was to dispel the mists of ornithological obscurity from the continent. 'I wish other men of science would show themselves as patriotic in their views as yourself,' he wrote to Gould in 1839. 'We know nothing of Australia as yet except that it is the land of convicts and kangaroos, & this after 50 years of possession! Why the French would have had 50 books, folios & sets published by this time. Indeed foreigners Germans [sic] are now anticipating us & we shall learn from them to understand the production of our own country, always excepting the ornithology.'

Others were not so convinced that John Gould, 'a mere museum man', was up to the challenge. Audubon, who had considerable experience of the hardships of field ornithology in the bayous and forests of Louisiana, was particularly sceptical about Gould's ability to survive in the bush. He told Bachman:

'Mr. Gould the author of the Birds of Europe is about

leaving this country for New Holland, or as it is now called Australia. – he takes his Wife and Bairns with him, a Waggon the size of a Squatters Cabin and all such apparatus as will imcumber [sic] him not a little – he has never travelled in the Woods, never salted his rump Stakes [sic] with Gun Powder and how he will take to it, will be a 'sin to Crockett'.

Undaunted, however, by these prospects, Gould set about the preparations for his journey with systematic thoroughness, and, considering the multitude of decisions that had to be made, remarkable clearheadedness. In particular, his decision to employ John Gilbert as his assistant in Australia was a masterly stroke of forward planning. In two years Gould could not possibly hope to cover all the areas he guessed would prove rewarding. He needed a trustworthy, dedicated, and professional assistant, whom he could dispatch to those parts he would not have time to visit personally. Once again, Gould's ability to pick his team was to prove invaluable.

John Gilbert came from Windsor and, like Gould, had studied as a gardener under John Townsend Aiton. Gilbert was eight years younger than Gould, but the two must have met earlier, for by 1828, at the age of 16, Gilbert was recruited by Gould to work as a taxidermist at the Zoological Society. The background of the two colleagues was remarkably similar, and Gilbert, with Gould's assistance, made considerable advances in his career. By 1835 the London Director had elevated both Gould and Gilbert from 'bird & beast stuffers' to 'naturalists'.

When Gilbert decided to leave London and apply for the post of curator at the museum of the newly established Natural History Society of Shrewsbury, Gould wrote a glowing letter of recommendation on his behalf. 'In consequence of the excellent character' he was given by Gould, Gilbert won the position; he also attracted the attention of one of the Society's honorary curators, the eminent ornithologist and anatomist Thomas

Left: A portrait of John Gould at the age of 45,
by T. H. Maguire.

Right: Elizabeth Gould, in a portrait
painted after her death at the age of 37.
The cockatiel she holds was
one of those brought back from Australia.

"Original rough design"
J. Gould

A rough sketch, attributed to John Gould,
of Temminck's sapphirewing (*Pterophanes temmincki*) for
A Monograph of the Trochilidae, or Family of Humming-birds.

Above: The only sketch verified to be by John Gould, illustrating his very basic skills as a draftsman. It was drawn during Gould's visit to America, at the Academy of Natural Sciences in Philadelphia.

Left: An early watercolor sketch by Elizabeth Gould for *A Century of Birds Hitherto Unfigured from the Himalaya Mountains.* Before she began to work under the instruction of Edward Lear, her drawings were somewhat frigid, often in silhouette, and clumsily composed.

The snowy owl (*Stryx nyctea*) by Edward Lear,
from *The Birds of Europe*.

Lear's plate of the culmenated toucan (*Ramphastos culmenatus*), from
Gould's *A Monograph of the Ramphastidae, or Family of Toucans.*

Edward Lear at the age of 28;
a pencil drawing by his friend W. R. Marstrand made in 1840,
a few years after Lear had left Gould's employ.

Charles Darwin at the age of 31;
a watercolor portrait by George Richmond
dated 1840.

The Australian goshawk (*Accipiter gentilis*), an original watercolor by Elizabeth Gould for *The Birds of Australia*.

Rosehill parakeets (*Platycerus ignitus*), a plate by H. C. Richter from *The Birds of Australia*.

Campbell Eyton. Gilbert began helping Eyton in the preparation of specimens for his private museum in his nearby ancestral home at Eyton, and, when Gilbert decided to leave England as Gould's assistant, Eyton offered to buy Gilbert's personal, 'ready stuffed' collection, including 82 birds and an 'almost perfect skeleton of an armadillo (with the exception of the toenails)'.

However successful Gilbert's career in Shropshire had been, it had not proved as spectacular as Gould's. When he resumed his post in Gould's employment in preparation for the expedition to Australia, it was in a position of marked subservience. Despite their common backgrounds, shared interests, and childhood ties, Gould's pre-eminence had swiftly distinguished him beyond social association with his former colleague. Gould assumed a superior status over his fellow naturalist, not only as his employer but as a new member of the middle-classes. The tone adopted towards Gilbert by the Gould company was one of patronising authority. In one of his letters to Gould in Australia, Prince writes, 'I and Joseph are both much pleased to hear so good a report of Gilbert. We trust he has seen his error and will become as respectable in his conduct as he is in his abilities.' Whatever Gould's personal opinion of Gilbert, or whatever his error, his qualifications as a naturalist were undisputed, and singled him out as the ideal candidate for sharing Gould's exploration. Moreover, Gilbert's eight years as an animal preserver under Gould at the Zoological Society had familiarised him with Gould's methods, and established a useful channel of communication between the two in the process.

Gilbert's commitment to the Australian project was as self-sacrificing as his employer's. He had married sometime in 1837 and had one child. Mrs Gilbert, it appears, died shortly before or during Gilbert's visit to Australia. Gould's arrangements for his own children were meticulously considered, which must have been of some consolation to his wife, who had dutifully resolved to accompany her husband to Australia. She was to leave behind three of the youngest children, the last a 'sickly babe' born less than six months before their departure; they were to take with them only their eldest son, John Henry, who was by then seven years old, and Mrs Gould's nephew, Henry Coxen, whom the couple had taken on when his father died in 1825. Their four-year-old son, Charlie, was sent away to school, while Lizzie, aged two, and the baby remained at Broad Street under the care of their grandmother Mrs Coxen. During her two-year absence Eliza suffered great pangs of homesickness and longing for her children. The distress of leaving them no doubt contributed to the 'sudden & severe indisposition of Mrs Gould . . . which inducing the utmost fears for her safety, rendered it very doubtful up to the last moment whether they would be able to go or not . . .'

Mrs Coxen was aided in her task of looking after the children by her niece, Mrs Mitchell, who also moved into Broad Street with her husband, who, in turn, was charged with helping to run the house. Together with Gould's uncle and a William John Martin of Camden Town, Mr Mitchell and Prince were appointed by Gould to act as his attorneys. Every effort was made by Gould to ensure that his business would run smoothly, efficiently, and with unremitting industry while he was away. The Gould machine was set like clockwork to continue ticking as methodically for the next two years as it had for the last eight. And it was on Prince's shoulders that this great burden of sustaining the Gould enterprise was to fall, as Gould ordained:

'It is my wish that my Clerk Mr. Prince who is well acquainted with my business and affairs and in whom I place great confidence should continue in the management and conduct of the same . . . to sell and dispose or exchange all or any of my works on Ornithology and specimens of Natural History in the manner I have been accustomed to do . . . to continue or complete as far as practicable the publication of any work or works of mine on Ornithology and to do all other [illegible] by issuing a Prospectus advertising the same . . . to purchase all necessary materials articles and things fit and proper for the carrying on of my business . . . to borrow for a temporary period any money from my Bankers, Messrs Drummond and Company . . . and if there by any surplus available for the purpose to invest the same in purchase of Stock . . . to pay the rent and taxes . . . make up, adjust and settle all and every or any Accounts. . . . [and generally] to do perform and execute all and every or any other acts deeds matters and things whatsoever are necessary to be done in all other my concerns engagements affairs and business whatsoever during my absence from England as fully and effectually to all intents and purposes as I myself might or could do if I were personally present and did the same.'

Although Gould had wound up two of his principal publications by the time he left England – the concluding part 22 of *Birds of Europe* was scheduled for July 1837, and the third and final part of the *Trogons* appeared on or before 14 March 1838 – Prince was charged with seeing to the publicity and the production of the plates for Darwin's *Zoology of the Beagle*, and the printing and colouring of the illustrations for the second part of *Icones Avium* on the species of. *Caprimulgidae*, or goatsuckers.

One has some sympathy for the loyal, compliant Mr Prince in these days before the Goulds' departure for Australia; even more so than when Prince was left behind with all the doubts and pressures of having to act on his master's behalf. If Gould could be hyperactive and despotic at the best of times, how much more demanding must he have been in preparing for this journey of a lifetime. Until the last minute Gould was dealing orders nineteen-to-the-dozen. Six months after Gould had left,

Prince had not cleared all the 'Memoranda taken while on board the Parsee'.

Somehow between them, however, they organised letters of introduction to naval captains who were to give Gould and Gilbert free passage whenever possible, to 'persons of influence' in general; and to the Governors of South Australia, Western Australia, New South Wales and Van Diemen's Land (now Tasmania). They assembled dissecting instruments, guns, collecting boxes, preserving bottles, spirits and drying agents; magnifying instruments, clothes and camping equipment; writing and drawing materials; arranged bank drafts, insurance and bills of lading; and engaged two personal servants for the trip.

When the *Birds of Australia* was finally completed in June 1848, Gould generously and justly acknowledged the vital role his amanuensis had played in running the entire business while he was away. Prince, he said, 'has been with me from the commencement of my various works. I left him in charge of the whole of my affairs during my absence from England, with a perfect conviction that he would zealously exert himself for my interest, and the confidence I reposed in him has been fully realised, not only during my absence, but during the long period of eighteen years.'

Although Gould might have increased his funds through sponsorship, he wisely insisted on remaining independent of any institutions, reserving the rewards of his labours for no one but himself. He had, according to Bowdler Sharpe, amassed a fortune of £7,000 from his publications by the time he left for Australia. For anyone else this might have seemed ample funds for an expedition, but for Gould, the cautious man of commerce, even this nest-egg had to be used sparingly. He had, after all, the business and a family to finance back in Broad Street, the cost of amassing and exporting his collections from Australia, and the travel and living expenses of seven people to cover. Even so, he resisted the temptation of applying for bursaries. Gilbert in particular was to suffer from Gould's overzealous housekeeping while they were in Australia, but the sacrifice was to prove, for Gould at least, worth every hardship. Gould resigned from the service of the Zoological Society but agreed to act as a 'corresponding Member', so that his letters and discoveries could be read aloud at the meetings and published in the Proceedings. In this way he would, much as Darwin had during the voyage of the *Beagle*, remain in the public eye all the time he was away.

Quite apart from his own necessities, Gould was beleaguered by requests from acquaintances to deliver letters to people in Sydney, even to trace a young man who had gone to Australia three years earlier and had not been heard of for more than a year. And of course he was beset by orders and suggestions from other collectors: William Ogilby advised him to collect 6 or 8 specimens of every species, particularly the great red woolly kangaroo, which were in great demand not only in England but on the Continent as well; Jardine asked for seeds as well as numerous other specified 'things in spirits'; Thomas Campbell Eyton wanted as many petrels as Gould could capture for a monograph on the *Procellaridae*.

On 30 April 1838 Gould wrote his last letter to Jardine before he left:

'I am so busy that my brain is in a complete jumble. I will however detail a few of my plans. In the first place I am happy to say I have a most comfortable ship the "Parsee" although a small one 350 tons Barque with a commodious poop, no steerage passengers and not more than ten or twelve cabin p[assengers]. Messrs. Jordan the owner and the Captain both take an interest in my presents [sic]. The Captain's name is I. McKeller and the mates Murray so you see we cannot do without your countrymen and candidly I shall place my life in their hands with the utmost confidence. I have a Stern Cabin eleven feet by twelve principally for the comfort of Mrs. G. and one adjoining of a smaller size. I have had the former fitted up with Drawers for birds and books in hope to work going along and taking a quantity of birds with me as well as left over plates to cut up for box books etc.

My party consists of Mrs. G. with an attendant a woman who has accompanied Ladies twice to Calcutta, our little boy, two men one as a collector, the other as a servant, and a nephew 14 years old who will settle there; in the next cabin is a Surgeon with his wife and family. You will say we have every comfort necessary to render so long and tedious a voyage as agreeable [sic] as it can be. We start in ten days and I firmly rely on the goodness of Providence to grant me a safe and prosperous passage. If permitted to return in safety I shall never regret the steps I am about to take. The voyage will certainly be a very expensive one but I doubt not my work will have a great sale in consequence and if so amply repay me for my exertions.

We go direct to Hobart Town. I have many reasons for visiting Van Diemens Land first – the climate is more like that of England and many of the Birds, I believe only pigeons may visit that island from the Continent, many are local species and others so nearly allied that [only] an inspection of them in a state of nature is possible to ascertain their being specifically different. From V. D. L. I either go to Port Philip or Sydney and from there into the interior. Mrs. Gould's brothers have estates high up in the Hunter just under the Liverpool Range and within five miles of the Menuras. I then if possible send my assistant to the north of the Island and Swan River. New Zealand I hope to visit myself probably on my way home as we come round the Horn and thereby circumnavigate the globe!!! What an age.

May yourself, lady J., and family every blessing and happiness, and believe me to remain

     Yours very Sincerely

     John Gould'

On 16 May 1838 John Gould and his party set sail from London in the *Parsee*. From the moment they left, Gould was in exuberant heart. Most would have dreaded the long, monotonous, uncomfortable journey ahead but Gould saw even the sea passage as an opportunity. The ship was constantly visited by birds, and as it ventured further and further from home shores, Gould's hopes of seeing new species, or at least species he had never seen alive before, rose to dizzying heights. A month into the voyage, Gould wrote to the Chairman of the Scientific Committee of the Zoological Society, in the highest of spirits:

> Latt 14-21. N. Long. 22-16. W
> June 20th 1838

My Dear Sir

I feel assured that you as well as many of my scientific friends will be happy to hear that we are thus far safely on our Voyage and that the whole of our party are quite well. We are now 7 degrees within the tropics and if the trade winds in which we got off Madeira continue we may expect to cross the equator in a week or ten days, about which time our Captn. informs us that we may expect to fall in with some homeward bound vessels which will enable me as well as the rest of our passengers to forward letters to our friends . . . bad weather had the effect of making us all good sailors and now we are enjoying fine weather and a fair wind. The past is quite forgotten and we are all cheerful and happy. During the bad weather we experienced a few disasters and events, but they were of little consequence. Our head rails were washed away and our water casks drove en masse from side to midships. I must say that I do not at present find the time at all tedious and I may say scarcely monotonous. In fact there has been so much to amuse that I have scarcely read a book and have written but little. There is always some new object presenting itself to view. Land birds are constantly visiting the ship and the pretty Storm Petrels are our constant companions while the ocean itself contains an abundance to amuse and gratify one. The whales of which we saw several in Bay together with grampuses and porpoises are all novelties. The beautiful floating Moluscas (Portugese Men of War!) and the luminous species which bispangle [sic] the water at night are to a lover of Nature all objects of interest, not forgetting the most beautiful of all a Shoal of Flying Fish.'

His letter contained an account of all the birds he had seen: a common goatsucker that flew gracefully round and round the ship for an hour, although it had 'in all probability passed the night on the wing'; a female yellow wagtail that alighted only for a moment; numerous petrels, five of which he killed for himself, and one for use in Mr Yarrell's work (a *History of British Birds*); a flight of swallows, and several small turtle doves, which 'visited the ship [and] went off again immediately'; a kestrel which was killed 'from off the Rigging'; a short-eared owl which flew on board during the night, was caught and kept alive for several days; and hundreds of shearwaters, which surrounded the ship off Madeira and the Salvage Rocks. At times, he wrote, the ship had been besieged by grampuses and a large shoal of black whales, each about 50 feet long.

Gould's letter to Prince, dispatched shortly after a two-day jaunt into the interior of the island of Tenerife, was apparently no less exhilarating. 'He writes in high spirits', Prince told Jardine, 'and appeared to be thoroughly enjoying this "new era of his existence".' On 18 August Prince replied to Gould's first letter:

'. . . we are delighted to find that the Voyage is productive of a high state of enjoyment to yourself and likely to be the means of re-establishing Mrs. Gould's health. How I envy you your excursion up the mountains of Teneriffe. The delight you must experience in the constant change of scene and the daily occurrences of so many objects of the highest interest is beyond our conception. . . . Mr Darwin asks if you "caught any fish?" . . .'

The voyage continued on from Tenerife towards the equator, from which point Gould began his researches in earnest. In another letter read out to the Zoological Society in his absence, but which was written on 10 May 1839, he recalls:

'We crossed the equator on the 7th of July, having been more than twenty days within the tropics, part of which time our vessel lay becalmed. This portion of the ocean's surface was also inhabited by storm petrels, but of a distinct species from any I had hitherto observed, and which I believe to be new to science. These birds, with now and then a solitary *Rhynchops* and frigate bird (*Tachypeles aquilus*), were all of the feathered race that I observed in these heated latitudes, a part of the voyage which always hang heavily upon those destined to visit these distant regions; by me, however, it was not so much felt, the monotony being relieved by the occasional occurrence of a whale, whose huge body rolled lazily by; by a shoal of porpoises, which sometimes perform most amusing evolutions, throwing themselves completely out of the water, or gliding through it with astonishing velocity; or by the occasional flight of the beautiful flying fish, when endeavouring to escape from the impetuous rush of the bonito or albacore.

'On the 20th July we reached the 26th degree of S. Lat., and were visited for the first time by the Cape petrel. On the 23rd, lat 31 S., long. 24 W., we found ourselves in seas literally teeming with the feathered race. Independently of an abundance of Cape petrels, two other species and three kinds of albatrosses were observed around us . . . A few days after this we commenced running down our longitude, and from this time until we reached the shores of Van Diemen's Land, several species of this family (Procellaridae) were daily in company with the ship. Whenever a favourable opportunity offered, Captain McKellar obligingly allowed me the use of a boat, and by this means enabled me to collect nearly all the species of this interesting family that we fell in with. . . .'

This was one of the ornithologist's happiest occupations during the sea voyage. 'My readers will easily imagine,' he wrote in his *Handbook to the Birds of Australia*, 'with what pleasure I descended the ship's side and sallied forth in a little "dingy" to procure specimens.' Nearly 30 species of oceanic birds were obtained by Gould in this manner on the way to Australia.

When the weather was too rough for a good shot, or for Gould to be lowered in a boat to collect his prize, he would stand on deck and fish for his birds with a hook and line. It was in this manner that, on one extraordinary and memorable occasion, he managed eventually to capture the soft-plumaged petrel:

'It is a species which will ever live in my memory, from its being the first large petrel I saw after crossing the line, and from a somewhat curious incident that then occurred. The weather being too boisterous to admit of a boat being lowered, I endeavoured to capture the bird with a hook and line, and the ordinary sea-hooks being too large for the purpose, I was in the act of selecting a hook from my stock of salmon-flies, when a sudden gust of wind blew my hooks, and a piece of parchment ten inches long by six inches wide on which they were lying, overboard into the sea, and I was obliged to give up the attempt for that day; on the next I succeeded in capturing the bird with a hook baited with fat, and the reader may judge of my surprise when on opening the stomach I there found the piece of parchment, so completely uninjured that it was dried and again restored to its original use.'

Not all the birds that Gould caught in this way were subjected to an autopsy. Species that were numerous around the boat and which were caught for a cursory examination only were released again alive, 'the operation', Gould maintained, giving 'not the least pain to the bird, the point of the hook merely taking hold in the horney [sic] and insensible tip of the beak.' The black-eyebrowed albatross was a frequent visitor to the *Parsee*, and gave Gould the chance to try out a crude but innovative experiment – one of the first attempts to tag and monitor a live sea bird. He 'caught numerous examples, marked and gave them their liberty, in order to ascertain whether the individuals which were flying round the ship at nightfall, were the same that were similarly engaged at daylight in the morning after a night's run of 120 miles, and which in nearly every instance proved to be the case.'

Altogether, during the outward voyage and the return passage around Cape Horn, Gould, as he described in his introduction to the *Birds of Australia*, 'was enabled to obtain nearly forty species of Petrel, being the finest collection of *Procellaridae* ever brought together', as well as examining and preserving numerous other species that came by chance to hand. Those which pertained to his projected work on the *Birds of Australia* he kept, about 20 species in all, surrendering the rest to his friend Eyton, as promised.

Some four months after leaving England, on 18 September 1838 the *Parsee* docked at Hobart Town, the picturesque capital of Van Diemen's Land. 'Mr and Mrs Gould are now in the Colony' proclaimed the *Hobart Town Courier*, 'to which they have come at great expense and sacrifice of comfort, purely with the view of making this work [*Birds of Australia*] still more valuable by taking their drawings from living specimens.' 'Now,' as the naturalist Tomasso Salvadori commented after his friend's death, 'arrived the most important moment in Gould's life, and the consequences which it had have never been equalled in the annals of ornithology.'

CHAPTER 7

The Harvest Meets the Sickle

B Y THE TIME of Gould's arrival, Hobart Town, which had been founded in the year of Gould's birth, boasted nearly 15,000 inhabitants. The area's particular attraction was, as Gould had anticipated, its superficial resemblance to England's green and pleasant land. Familiar names such as Ilfracombe, Dorset, Devonport, Tamar, Launceston, Bridport, Somerset, and Westbury echoed the geography of the West Country, and the estuary on which Hobart was founded was reassuringly called the Derwent. Houses that lined the wide streets were made of brick and roofed in shingle to imitate the appearance of slate. Each stood separately in its own little suburban garden. Above the town rose the 4,000-foot peak of Mount Wellington, the only vestige of untamed nature. Around the bay the hills had been cleared, and patches of emerald fields punctuated by tiny white-washed cottages carried the wistful eye back to the countryside of Ireland. The sea stretched for miles from the prospect of Hobart and was populated at closer quarters by the comforting bustle of shipping.

The Gould party settled into accommodation at a Mr

Fisher's on Davey Street, unpacked all their belongings and set up their apparatus ready for the invasion of fresh specimens. Gould lost no time venturing into the field, taking with him on some excursions his nephew Henry, who was understandably thrilled by the prospect of driving bullock carts; his servant James, whom he was training in the art of taxidermy; and sometimes his assistant Gilbert. Gould was helped in his researches by local naturalists like the Reverend Ewing, after whom he later named the native Tasmanian fruit pidgeon and a species of *Acanthiza*.

Soon Gould was immersed in the business of collecting. He did not have far to go to find an enthralling abundance of birds. Tree swallows had just arrived in Hobart on their annual migrations, and were 'particularly numerous in the streets'; wood swallows dived among the paddocks and pasturelands on nearby estates; fire-tailed finches and flame-breasted robins were busy nest-building in gardens and orchards – 'I have even taken its nest', remarked Gould about the latter, 'from a shelving bank in the streets of Hobart' – while dusky robins perched on garden railings. In the parks the eucalypti were alive with lorikeets, and flocks of parakeets chased each other around the streets; while spotted diamond birds clung to the leaves of trees in every enclosure, and yellow-throated honey-eaters crept along branches in the ravines around the town. Most of Gould's collecting in Van Diemen's Land, however, was done within 80 miles of Hobart, on or around the Derwent River and the fertile Macquairie Plains, where kingfishers, water crakes and crow-shrikes, golden plovers and white-fronted herons, bitterns and black-backed porphyrios were to be found in glorious profusion.

For Elizabeth Gould, however, this ornithological cornu-copia right on the doorstep did not hold the same all-encompassing fascination. She felt the distance from England far more acutely than her preoccupied spouse; her thoughts were with the children, and, even at this early stage, she longed for the voyage home. In the first letter to her mother that reached England, Eliza wrote home with mixed emotions of excitement and homesickness.

Hobarton. Oct. 8th 1838

My dear Mother,

We arrived safely on the 19th [18 September according to the *Hobart Town Courier*] of September in excellent health, and but for the thoughts of those we left behind should also be in good spirits as our prospects here are in many respects cheering. The country is very fine, teeming with beautiful natural productions, both in the animal and the vegetable kingdom. Persons to whom we have been introduced are exceedingly kind and John is acquiring a vast fund of information in the ornithological department, which must, I think, prove interesting to the lovers of that science.

We got here just in the right season, and I assure you he has already shown himself a great enemy to the feathered tribe, having shot a great many beautiful birds and robbed various others of their nests and eggs. Indeed John is so enthusiastic that one cannot be with him without catching some of his zeal in the cause, and I cannot regret our coming, though looking anxiously forward to our return. Could I be sure of meeting you all again in health I could be content, but there is the anxiety . . .

. . . I trust we shall be back in two years from the time we left England . . .

Before long, though, Eliza's melancholia was relieved by the couple's introduction to Sir John and Lady Franklin. Sir John Franklin, who had been dubbed the 'Polar Knight' for his valiant exploration of the Arctic, had been appointed lieutenant-governor of Van Diemen's Land the previous year. It was to be another seven years before Franklin returned to his first love as commander of the *Erebus* on his most famous – and fatal – mission to find the legendary north-west passage from the polar seas to the north Pacific; until then he occupied himself with the social and moral improvement of the colony in his charge. Sir John was deeply involved in reforming the conditions and prospects of the island's large convict population, and greatly interested in exploring the land and its resources. He was particularly keen on the development of culture in the young colony. Franklin, who was 18 years Gould's senior, had established 'for the better class of colonists' a scientific society, the first of its kind in the colonies (and later to become the Royal Society of Hobart Town) and welcomed the ornithologist's arrival with great enthusiasm. Gould had been given a number of letters of introduction to the veteran explorer from, among others, George Bank, who had accompanied Franklin on one of his Arctic expeditions, and Captain John Washington, secretary of the Royal Geographical Society.

Friendship rapidly developed between Mrs Gould, who was pregnant for the seventh time, and Lady Franklin, who at 46 had no children and was delighted to entertain a young family. Lady Franklin, a 'lioness of a woman' with a lively curiosity and intelligence, devoted much of her time to promoting the visual arts in the infant colony, and took a keen interest in Mrs Gould's work for her husband. She was as much a natural history enthusiast as her husband, and had founded a botanical garden on the banks of the Derwent just outside Hobart. Eliza surely spent hours scouring the beds and walkways for native plants to use in her bird drawings.

In Eliza's fourth letter home to her mother (the third was lost) in which she again expresses her yearning for the children, she also describes her developing attachment to the Franklins, and her husband's first significant expedition under their auspices:

'I feel extremely anxious about you, being well aware that

this time of year is most trying in England. I sometimes picture to myself both you and Mrs. Mitchell suffering from the winter colds and fogs. My dear little Louisa too is just at a critical age, teething in all probability. I did not forget the darling's birthday. Bless their dear little faces. How I love to recall their looks to my mind's eye . . .

. . . Now for the most important part of my very rambling letter. As John would say, that is about the birds. In the first place he has desired me to say he would write but for his constant occupation. He is now out and as soon as he returns is to accompany Lady Franklin and her party in a vessel round the island as far as Port Davey and Macquarie Harbour, which you may easily see on the map. I was to have gone also but declined. John hopes to make there considerable additions to his stores. He has been already very successful both as regards birds and their nests and eggs; he has a beautiful collection of them. Before leaving V. D.'s L. he intends sending home a large case of skins, skeletons, eggs, etc.

All persons (or nearly so) to whom we have been introduced have endeavoured to promote our views, and here I must not forget to mention the great kindness we have received from Sir John and Lady Franklin in every way. We have been frequent visitors at Government House and have been staying also at their cottage, New Norfolk. I have also been invited to take up my abode at Government House during Lady F's absence at Macquarie. Sir John, his niece, and some other ladies staying there, Lady F. thought it would be less lonely for me being in lodgings solitary . . .'

The indomitable Lady Franklin took off with Eliza's husband around the 10 December on his first major exploration of the south and west coasts of Van Diemen's Land. They left Hobart with a small party on a government schooner bound for Port

Davey, but got no further than Recherche Bay before being prevented by stormy weather for a fortnight. Lady Franklin kept a journal during the voyage in which she described Gould's attempts to continue bird collecting while they waited, windbound, in the Bay:

Tuesday, December 11th. Entrance to D'Entrecasteaux's Channel.

'About 2 o'clock, when within about 4 miles of Green Island, for which we were tacking, Mr. Gould went off in a boat with the hope of reaching it before us, and of finding some penguin's eggs which he is much in want off [sic]. He had landed on the island before and killed penguins, quails, ducks, etc. On this occasion he did not succeed in finding any penguin's eggs, but came back with a live penguin, and with the eggs of various gulls. Some of these were too hard to be blown, in which case he cut with the point of a penknife or of a small knife adapted for the purpose, an oval-shaped piece of the shell out of the side, emptied the egg, and replaced the shell.'

December 12 Research [sic] Bay.

'Mr. Gould went off in the morning to the Acteon [sic] Islands, which are about a mile distant from each other and at the distance of about 3 or 4 miles from Research Bay. The soil of them is sandy and much covered with scrub. He procured there 2 species of parrots he had not yet taken, an albatross, teal, gulls, and the eggs of the latter.'

December 13.

'Messrs. Gould and Gunn set off to-day to the head of the [Recherche] Bay to a plain, which appears to run up many miles into the country the hill called South Cape, which presents towards the bay a steep and particularly denuded surface. . . . Mr. Gould expected a rich harvest from the appearance of the plain, both as respect [sic] quadrupeds and birds, but it was remarkably destitute. He brought back, however, the nest of an emu wren, and a parrot which he had not killed before.'

The weather did not relent. The entire party, even the persistent Mr Gould, was forced to spend an entire day on board ship. Two days passed, and on Sunday, 16 December, the schooner tried for a second time to get to Port Davey but was forced back. Gould was getting impatient. Lady Franklin wrote: 'Monday, December 17th. It blew very hard, but not so much as to prevent Messrs. Gould and Gunn going on shore. They visited the stream called Catamaran River in the Port au Nord.'

The following day Gould visited the Catamaran River area again, but by now he was running short of fresh collecting sites and desperate to get back to work in earnest. By Thursday he was considering aborting the venture altogether and returning to Hobart.

Thursday, December 20th.

'Mr Gould was worn out by our reverses, regretted the loss of time, and this very afternoon had been declaring with many apologies that he must go back in the *Vansittart*. His good humour under his prolonged disappointment had never failed, but his time was precious to him. When he came to V. D. L. he intended to stop only a month. How should he get through his work if he went on in this way? Mr. Gould was to my very great regret in this disposition when Mr. Gunn, Elinor and I embarked again in the evening to visit the shores of that portion of the northern part of Research Bay where La Haye's garden was planted, and where Captain King had discovered signs of coal . . . On our return to the schooner we found Mr. Gould no longer firm in his former determination to depart immediately, and on my telling him he must give me his hand in pledge that he would stay here and work longer if necessary, he, after a little hesitation, consented. As a little compensation I begged Captain King to let us remove either to Bruny, or to Muscle Bay, which would make very little difference when once the wind set fair, and it was accordingly settled that at daylight we should sail for Muscle Bay.'

For once, Gould's impatience was overruled. The iron will of Lady Franklin bent the egotistical ornithologist under her sway, and he was forced to capitulate to someone else's wishes. Despite the promise of great ornithological rewards if he stayed, Gould found it immensely difficult to remain inactive while there was so much work to be done. Time was weighing heavily upon him. That evening Gould wrote to his wife describing his frustration at having to wait yet longer in Recherche Bay:

Recherche Bay   Thursday Night
20 Decr 38

Mr Dear Eliza,

Your letter of Sunday last was put into my hands this morning it having taken the Pilot with 5 men 4 days to reach this place so adverse has been the weather, you will therefore see we are still imprizoned [sic] here. We endeavored to get round the South West Cape on Sunday morning but was obliged [sic] quickly to retrace our steps. The Cutter Vansittart is here also and will take this letter to town. I would gladly have availed myself of the opportunity of taking a passage in her myself did I not find by so doing that the party would be entirely broken up and in all probability the trip to the westward abandoned altogether. I myself would I am sure hereafter regret not visiting this part of V. D. L., and under all circumstances

I have agreed with Lady Franklin to wait one week longer after which to return if the wind should not come round to the eastward or northeast, either of these winds would take us to the desired place in a few hours.

We intend shifting our locality tomorrow morning; we are going into South Port 15 miles distance from this, (nearer home) this move is principally on my account for the purpose of affording me a new locality for my rambles.

I have this day killed rather an extraordinary Bird, and one that you will recollect, at least there is every probability of its being the same; you will remember the large Snow White Petrel (a variety of the black species) which followed us nearly all the way from the Cape to Tasman's Head and which I was so desirous of procuring without effecting my purpose.

I have also killed some beautiful albatrosses, petrels, ducks etc. and four nests with full compliment of eggs of the Black Oystercatcher, small and large gulls, a Teal etc. besides a nest with young of the Emu Wren and other minor things.

I am quite well and am happy to say Lady F. is much better and is quite well. I sincerely trust you are so also and that you are finally established at Government House. Tell Henry I hope he is good and very attentive to you and it is and will be his place to be your protector! in my absence. I am sorry you have not heard from Sydney, you cannot be long without letters, if they are not arrived already [letter torn]. The Marian Watson will return immediately. This is a cold cheerless place and a complete charnel house for whaling. The beach is almost everywhere strewn with their Bones, entire skeletons many of which are still putrid. You may therefore guess what we find to leeward.

I think I have not more to say.

God bless you my dear Eliza.

J. Gould

Hastily

    P.S. I think we left some skeletons in water, if so, tell James to take them out merely and lay them on the table to dry.

They decided to move on to Muscle Bay, which provided Gould with few prizes. Together with Thomas Smith, assistant police magistrate at South Port, he explored the bay and an inlet and was later able to report 'I procured newly-hatched young [of the Black Swan] clothed in greyish white down at South Port River'. But scarcely had he begun to investigate these new, if somewhat less adventurous, hunting-grounds, than the entire party was 'summoned back to Hobarton by Sir John'. The expedition returned only two weeks after it had left, having never got to see Macquairie, the south coast, or the promising territory around Port Davy.

But the venture had not been a failure. Gould had procured, as Elizabeth reported to her mother, 'several species of interesting birds while in Recherche Bay and has now I believe most of the birds of this colony.' Gould's impatience and his disappointment stemmed not so much from the fruits of his collecting, but from the frustration of inactivity and the miscarriage of a good project. He set off again almost as soon as he had returned to Hobart to make up for lost time on an overland expedition to Launceston and George Town, 60 miles from Hobart on the North Coast of the island.

By then the Goulds had received their first batch of letters. They had been sent on from Sydney by their nephew Henry, who had left for Australia a few weeks earlier to live with his uncle, Stephen Coxen, at Yarrundi in New South Wales. Among the letters were the first three massive instalments from Prince detailing all the business transactions to date, a letter from Mrs Mitchell, and one for Eliza from Mrs Coxen. On 3 January, Eliza sat down to reply to her mother:

'. . . Doubtless you are surprised at our being so long here. John has found much to do which will not require to be again done in N.S.W. I still hope we may not exceed two years in our absence and I think John is as anxious as myself on the subject. Meantime it gives us great consolation to know that you and our dear children are well. Do not let them quite forget Momma and Papa if possible. I sincerely wish you many happy returns of the season which I can scarcely realise as the winter of England, so odd does it seem to have the thermometer at 99 on Xmas day.

I am so very glad to hear that my darlings are all making such splendid improvement. Do you mean little Louisa's propensity for breaking plates as an example? Bless her sweet little face! Who does she grow like? And my dear little Eliza, is she as pretty as she was? You may laugh, but she was pretty, at least in my eyes. And dear good-tempered little Charley . . .

And now for ourselves and movements, as that is a welcome thing to you I know. . . . John on Tuesday last set out with Gilbert for Launceston, intending to visit some island in Bass's Strait. He will be gone about a month and on his return we shall go to Sydney as quickly as we can. . . . I am during his absence drawing as many native plants as I can, I mean branches of trees, some of which are very pretty.

Henry is staying with me at Government House at present. Sir John and Lady Franklin are very kind and wish me to feel quite at home. . . . [John] wished me to say for him everything that was kind to all, especially to his mother, sisters, and Mrs. Cleave and Mrs. Stuart, also Mr. and Mrs. Mitchell. He is extremely occupied. His not writing more frequently is really excuseable, as you would acknowledge could you follow his movements as he slaves about all day in untiring perseverance . . .'

In a postscript Eliza somewhat enigmatically mentions her pregnancy for the first time:

> 'My dear Mother, Mrs Mitchell says, "Do let us know if you mean to bring home a little convict." I did not mean to say but perhaps it is better I should, therefore only remark that I will speak on the subject in my next, but do not suffer alarms on my account – I am in better health than I have been for years and I hope to get to Sydney and get comfortably settled in good time. . . .'

Gould meanwhile was some 60 miles north of Hobart in the 'interior of Van Diemen's Land' at Spring Hill Inn, his thoughts engrossed in collecting. Audubon's scepticism about Gould's ability to meet the strenuous physical demands of life in the bush proved to be without foundation. Gould rose to the occasion and, driven by the feverish excitement of a treasure-seeker, surmounted every obstacle Tasmania's chequered geography could throw at him. On 3 January he shot what he later called the 'Clinking Currawong', or *Strepera arguta*. With somewhat unconvincing ostentation, given Gould's ignorance of Latin, he describes how he coined a name for this unusual species. 'Its note is a loud ringing and very peculiar sound, somewhat resembling the words *Clink, clink*, several times repeated, and strongly reminded me of the distant sound of the stroke on a blacksmith's anvil, and hence the term "arguta" appeared to me to be an appropriate specific appellation for this new species.' Might this, one wonders, be one of the names that Albert Gunther so helpfully supplied Gould with over a chop at the British Museum?

Blue-banded grass parakeets abounded in the neighbour-hood of Spring Hill, feeding on corn from the margins of cultivated land in flocks larger than Gould had ever seen before; compared with the proliferation of different species around Hobart, however, only a few birds were collected on this overland journey. The main focus of Gould's attention was the fecund islands to the north-west of Tasmania in Bass's Straits, whence they would travel by ship from George Town. After a brief stay in Launceston on the upper reaches of the Tamar, where he replenished his stock of shot, powder, and caps, Gould wrote to his wife in Hobart:

> George Town, River Tamar
> Jany 8th 1839

My Dear Eliza,

    We reached this place about ten this morning and are at present wind bound but trust we shall be fairly under way for Flinders tomorrow morning. Your welcome letter found me at Launceston at which place I remained two days with Mr. Kerr, Captn Friend etc. I was not so much gratified with the interior of the country as I had anticipated but the people tell me I have seen it in an unfavorable season, in consequence of no rain having fallen for 3 months. The ground is literally parched and every blade of grass is yellow as sulphur, but independent of this the land in the interior is less rich and the fen of the country less fertile than any country I have ever traveled [sic]. On the whole the roads were good and also the Inns – the latter however are extravagantly high in their charges.

    I collected a few birds on the way, among them a splendid Eagle which I killed at Perth. This place George Town is something like I should think Thenness [?] when that town was first established although the extensive docks of the latter render it now far more different.

    I look forward with considerable pleasure to our arrival at Flinders whither I proceed purposely to see the remnants of the inhabitants which once peopled this fine Island over which they were Lords and Masters but now submissive creatures to the wiles of Englishmen. It will also be an interesting as well as expensive trip to me in other points. After Flinders you are aware I go to Kings and although I trust it will not be so it might take a month to accomplish the journey (that is if the winds prove contrary) I state this in order that you might not be alarmed in not hearing from me. You may depend if an opportunity should occur I will not fail to send a letter and I have to request in a [letter torn] fervor that you will post a letter to me at least once a *week* commencing from Monday next, with all particulars respecting your health etc etc directing them P. Office Launceston until called for, and when I again reach this place I will write immediately. I am happy to hear that Henry was drawing and trust he will be attentive to you in every way. I should say give Sophy a Dollar, and I will think of the men when I return.

I wish you to send by the Launceston coach two *clean* parts, of course, (nos. one and two) of the Birds of Australia with some prospectuses [intended for Gilbert, who was sailing directly to Swan River from Launceston, to take to John Hutt, Governor of Western Australia] which you will address to Captn Friend Post Office Launceston. You will either get the four nos of the Synopsis from the work room drawing portfolio, in Government House and send them with the larger copy.

You had better get Fisher to plain [sic] a very thin teak board the size of the large books for protection, and a common book cover from the work room will do for the protection of the other side, spreading the four parts of the Synopsis over and sealing the whole in brown paper. This parcel I wish *sent immediately*; James can pack it.

Must now conclude trusting my dear that every blessing will attend you and trusting I may find you well when I return – kiss dear Fritz [query] for me and believe me to remain Eliza

Yours ever affectionately

John Gould

The islands of Isabella, Green Island, Flinders, and Waterhouse were to prove a fertile hunting-ground even if not aesthetically pleasing. 'Nothing could be more sterile and parched than the islands in Bass's Straits', wrote Gould later, comparing them to the idyllic and verdant plains of the Upper Hunter River in Australia. His collecting tasks in Bass's Straits were not made easier, either, by the interruptions of the wattled plover, a particularly noisy bird, bent, it seemed, on frustrating all Gould's efforts. As Gould wrote in his handbook:

'In some parts of New South Wales this ornamental bird has obtained the name of the Alarm-bird from its rising in the air, flying round and screaming at the approach of an intruder, causing not only all of its own species to follow its example, but every other animal in the district to be on the alert. This fact I had ample opportunity of verifying on the islands of Bass's Straits, where I had scarcely stepped from the boat before every creature was made acquainted with my presence – no small annoyance to me, whose object was to secure the wary cereopsis and eagle, which with thousands of petrels and many other kinds of water-birds tenant these dreary islands.'

Birds were plentiful, but in many cases not as plentiful as Gould had been led to believe. The islands were the province of the 'sealers' and countless other opportunists intent on making their fortune from the area's abundant 'natural productions'. For the first time Gould came up against the devastating effects of commercial exploitation. The *Cereopsis*, or Cape Barren goose, had become an early victim, being prized for its downy feathers. 'This is one of the Australian birds', noted Gould in his Handbook, 'which particularly attracted the notice of the earlier voyagers to that country, by nearly every one of whom it is mentioned as being very plentiful on all the islands in Bass's Straits, and so tame that it might be easily knocked down with sticks or even captured by hand; during my sojourn in the country I visited many of the localities above mentioned, and found that, so far from being still numerous, it is almost extirpated; I killed a pair on Isabella Island, one of a small group near Flinder's Island, on the 12th January 1839.'

The following day, 13 January 1839, found Gould on Flinder's Island, where he took among other prizes, 'five newly laid eggs' of a black swan. The presence of the black swan in the islands only emphasised for Gould its unnatural absence from the large rivers, such as the Derwent, on mainland Tasmania where it once abounded:

'In the white man, the Black Swan finds an enemy so deadly, that in many parts where it was formerly numerous it has been almost, if not entirely, extirpated. . . . One most destructive mode by which vast numbers are destroyed is that of chasing the birds in a boat at the time they shed their primary quill-feathers, when being unable to fly they are soon rowed down and captured; this practice, which is to be much regretted, is usually resorted to for the sake of the beautiful down with which the breasts are clothed, but not unfrequently is mere wantonness. I have heard of the boats of a whaler entering an estuary and returning to the ship, nearly filled with Black Swans destroyed in this manner.'

The albatrosses too were mercilessly hunted and often left, plucked and dying, where the sealers found them. In the year before Gould's arrival a thousand albatrosses were killed on Albatross Island alone. Nothing, however, could compare with the wholesale carnage committed by men among the short-tailed, or sooty, petrel, known to the sealers as the mutton bird. Despite its relentless slaughter, its numbers were still great enough to astonish Gould when he visited one of its principal colonies on Green Island. He was still distressed, however, by the prolific trade in mutton bird feathers, two-and-a-half tons of which could be taken in a season – the produce of roughly 112,000 birds – and which would be sold for about 6d per pound in Launceston.

If Gould was disgusted by the commercial exploits of his fellow white men, he was compensated by the reverence for birds that he discovered among the Aborigines. His sojourn on Flinders Island became one of the highlights of his stay in Tasmania. In all probability he was greeted at the tiny settlement known as The Lagoons by a 'corrobery', just as the missionary James Backhouse had been the year before. The Aborigines had suffered at the hands of the sealers – many of the women were kidnapped and put to work plucking birds for the feather trade – but they retained their custom of a spontaneous welcome for anyone who put into port on the island. It is delightful to imagine Gould being invited in among their thatched huts to watch the famous emu dance.

Gould was enthralled by the spirit and customs of the Aborigines, and above all their almost religious respect for nature and its creatures. He was in many ways far ahead of his compatriots in his attitude to these original occupants of Tasmania. 'I shall always feel it a bounden duty on my part', Gould later wrote to Sir William Jardine, 'to sing their praises for if well treated they are the most harmless kind of creatures imaginable.'

Throughout his travels in the southern hemisphere Gould's behaviour towards the Aborigines is distinguished by a fearless, compassionate, and unbiased recognition of their plight, and a profound respect for their way of life. The interest in birds that he shared with them transcended the prejudices and the racism of resident colonials. In turn, many of the Aborigines Gould encountered, and especially those who accompanied him as guides on his collecting expeditions, grew to admire him and to recognise a kindred spirit in the eccentric and enthusiastic white man. Gould grew to depend on the Aborigines' local knowledge to find birds that were new to the Western world and to learn about their habits. With uncustomary generosity, he included in his letterpress the Aboriginal name for a species, when he knew it, alongside its Latin and common names.

On his exploration of the islands in Bass's Straits, Gould was assisted by a native servant lent him by Lady Franklin. 'Tell Lady Franklin,' Gould instructed his wife in a letter to Hobart, 'I have her little page with me, he is a most interesting little fellow, throws the spear and waddy with the utmost dexterity and [is] extremely useful to me in the bush, an eye like a hawk discovers birds nests & eggs in a most astonishing manner.'

Unfortunately, the voyage onward from Flinders' Island was curtailed by a horrifying incident that left Gould in a state of shock and unable to continue the expedition. He wrote to Eliza from George Town before commencing the journey overland back to Hobart, describing what had happened:

Sunday 20th Jany 1838 [1839]

My dear Eliza,

I am now in Capt. Friend's office and have only a short time to write before the post leaves. We arrived here yesterday from Flinders at which place I was especially gratified to [word not clear] my acquaintance with the Natives and other things, and I should have left the island with a light heart and proceeded to Kings had [not] a fatal accident happened to one of the men who shot himself dead by uncautiously pulling the gun from the boat with the muzzle toward his chest, the cock of the gun caught the seat of the boat and all was over with the poor fellow in half a minute. I cannot tell you my Dear Eliza how great a shock I sustained. I have scarcely been myself since and I almost hate the sight of a gun. I have given up all idea now of going to Kings and shall make my way across the Island as quickly as I can making a call or two on the way. The inquest will be held in the morning. The man had every caution given him not a minute before to be careful with the gun, but his time was come as his poor shipmates say and with that they console themselves. . . .

It must have been with some degree of cautiousness that Gould once more stocked up on shot, caps, and powder on his way back through Launceston for the journey home. It was in Launceston that Gould decided to part company with Gilbert, instructing him to await the arrival in Launceston of the *Comet*, which would carry him on to the Swan River in Western Australia. It was a typically *ad hoc* decision on Gould's part, and Gilbert was forced to continue his journey without his belongings, entrusting his trunk, which was back in Hobart, to his employer's safekeeping.

Back once more in the capital after an uneventful overland journey, Gould took a rare moment to sit down and recount his adventures to the household in Broad Street. Prince reported them faithfully to Sir William Jardine:

London 20 Broad Street, Golden Square
July 20, 1839

Sir,

Sir Philip Egerton having been kind enough to give me a Frank I embrace the opportunity to send you some account of Mr. Gould and his movements, presuming that as you expressed a wish to hear of him from time to time, a letter on this subject might not be devoid of interest.

Our last Letters are dated Hobart Town Feb. 9 up to which time his expedition had been eminently successful; far more so than he could have anticipated; the most liberal assistance had been rendered him by the Authorities, everything that could facilitate his views being cheerfully accorded: while nothing could exceed the kindness of Sir John and Lady Franklin in whose house he was then residing: in fact so much were they interested in his pursuits than upon more than one occasion they accompanied him in his exploring parties.

They have been over the greater part of Van Diemen's Land and the Islands in Bass's Straits and appear highly gratified with their peregrinations: a government vessel has generally been devoted to their use. The beauty and novelty of the scenery, the luxuriance of the shrubs and above all the originality of the Natives has astonished them beyond description, and so raised their enthusiasm that they seem scarcely to have felt the labour and fatigue of ascending high mountains or traversing deep glens and ravines, in fact so many wonders in the shape of animal creation have sprung up, as it were before them, that their imaginations have been kept in one continued state of delighted excitement. The result of his labours during the five months he had been there is a collection consisting of 500 Birds; 100 Skeletons, 60 nests, the eggs of alike [sic] No. of species, and 3 nine gallon Kegs of specimens in spirits . . .

. . . The whole of the party were in excellent health and had well tried each others stamina in walking up mountains etc. etc. but having so many interesting objects in view they had allowed no difficulty or danger to impede their progress.

Typically, Gould's infrequent letters home detailed only his ornithological exploits, much to the dismay of the household, who were desperate to hear news of a more personal nature. On 2 March 1839, Prince replied with a sense of disappointment to Gould's impersonal letters:

'On the 7th and 10th of Feb. we rec<u>d</u> your Letters of the 25th of Sept<u>r</u> and 6th of Oct: They were a source of the highest gratification to us all, but neither Letter was so full or explanatory as we could wish and from the slight mention of Mrs. Gould we are fearful that her health is not much improved. You do not enter into any details respecting the Voyage, where you are resident etc. in fact I gleaned more information on these points from one of Gilbert's Letters to a friend of his than from all 3 of yours – We are, however, delighted to find your expedition progressing so satisfactorily and anxiously hope it may continue to do so.'

Despite Prince's pleas, there was no improvement in the ornithologist's communications, and a few months later, in a letter dated 20 May, Prince took the opportunity to berate Gould for his thoughtlessness:

'. . . we are of course much gratified to hear of your safety and that your expedition is progressing so prosperously: at the same time the Letter although a long one does not tell us half the things we most wished to know: it does not say ought of your movements, when we may expect you home, how Mrs. Gould is, whether your attendants have answered your expectations, how Mastr. Henry is, or whether Mrs G. has presented you with a little Tasmanian, or likely to do so, a point upon which her mother is so exceedingly anxious that on finding it was not mentioned she sat down and cried with vexation.'

Elizabeth's letters were far more accommodating. While Gould was away in Bass's Straits she described how she occupied her time without him, painting plants and flowers for later use in the plates, and sketching the bird specimens delivered to her table. Her pregnancy thankfully gave her a little relief from the enormous output normally expected of her, and her stay with the Franklins was a welcome respite from the boredom and loneliness of being on her own, as she wrote on 9 January:

'We are very comfortable at Government House and Lady Franklin will not hear of our going from it so long as we remain in town. By *we* I mean Henry and I . . . Could you see us you would be astonished and I think pleased at the change of life this visit to lady F. affords me. They give many dinner parties but otherwise live in a very quiet, regular way. Do not, however, suppose that I am quite idle, though happily for my health it is not requisite to apply so closely as formerly was the case. *Wait a little, there will be enough of it* bye and bye and to spare.

Just now during John's absence I find amusement and employment in drawing some of the plants of the colony, which will help to render the work on Birds of Australia more interesting. All our sketches are much approved of and highly complimented by our friends. I wish you could hear some of the magnificent speeches that are frequently made us, because I know you like dearly to hear your daughter praised. But at the end of all I sigh and think if I could but see old England again, and the dear, dear treasures it contains, I would contentedly sit down at my working table and *stroke, stroke* away to the end of the chapter, that is health permitting.'

These were difficult times for Elizabeth. She more than anyone else understood her husband's obsession and his ambition to become Australia's first ornithologist; she encouraged and supported him. But she could never hope to share in his excitement and his passion to the same degree. The dream that had brought the two of them to the other side of the globe was wedging a distance between them. Eliza's priorities were as a mother, and though Gould was sympathetic, there was little room for compromise. Eliza was now six months pregnant, and Lady Franklin extended a 'pressing invitation' to her to remain at Government House for the rest of her confinement. Eliza acquiesced, forsaking for the time being the prospect of seeing her brothers in Australia for the comfort and security of staying in Hobart, and yet sad to be losing the company of her husband. 'John left late last night by the "Potentate" for Sydney, without me, much to my sorrow', she told her mother on 15 February 1839, 'though I cannot but admit the propriety of the arrangement.'

Gould left for Sydney, with plans to return in time for the birth, excited and full of anticipation at the prospect of 'fresh fields'. He had already exceeded his highest hopes, and he anticipated equal if not greater success from the great continent. On the day he left, Gould wrote to Captain Washington at the Royal Geographical Society, announcing proudly, 'I shall have many novelties to add to science as well as to communicate a great deal respecting their habits since I have already obtained the nests and eggs of 60 species not one of which as far as I am aware have been described.'

He had added two new candidates, the *Acanthiza ewingi* and the *Strepera arguta*, to the list of native Tasmanian species, making a total of 12, four of which he had already named from specimens sent to him before he left England, and had now succeeded in seeing all of these alive in their native habitat. Later, in 1855, he was to name a thirteenth Tasmanian species, the scrub tit or *Acanthornis magna*, from a specimen sent to him by his friend Ronald Gunn. Expecting similar rewards from the next leg of the expedition, Gould left Van Diemen's Land full of optimism and with a light heart.

CHAPTER 8

Adventure in the Newest
World

I F TASMANIA COULD BE REGARDED as a micro-
cosm of what to expect from the great continent of
Australia, Gould's exhilaration as he sailed towards
Sydney could not, understandably, have been higher. He was
not a superstitious man, but even he must have been
encouraged by the good omens that visited him while on board
the *Potentate*. The short, nine-day voyage was accompanied by
beautiful weather, and brought with it one of Gould's most
elusive species of petrel, one that had tantalised him for weeks
aboard the *Parsee*, although laying hands on it was, as Gould
liked to stress, down to his own ingenuity.

'While engaged in watching the movements of the several
species of the great family of *Procellaridae*, which at one
time often and often surrounded the ships that conveyed
me round the world, a bright speck would appear on the
distant horizon, and, gradually approaching nearer and
nearer, at length assumed the form of the Whiteheaded

Petrel, whose wing-powers far exceed those of any of its congeners; at one moment it would be rising high in the air, at the next sweeping comet-like through the flocks flying around; never, however, approaching the ship sufficiently near for a successful shot, and it was equally wary of avoiding the boat with which I was frequently favoured for the purpose of securing examples of other species; but, to make use of a familiar adage, "the most knowing are taken in at last"; one beautiful morning, the 20th of Feb. 1839, during my passage from Hobart Town to Sydney, when the sea was perfectly calm and of a glass smoothness, this wanderer of the ocean came in sight and approached within three hundred yards of the vessel; anxious to attract him still closer, so as to bring him within range, I thought of the following stratagem: – a corked bottle, attached to a long line, was thrown overboard and allowed to drift to the distance of forty or fifty yards, and kept there until the bird favoured us with another visit, while flying around in immense circles; at length his keen eye caught sight of the neck of the bottle (to which a bobbing motion was communicated by sudden jerks of the string), and he at once proceeded to examine more closely what it was that had arrested his attention; during this momentary pause the trigger was pulled, the boat lowered, and the bird was soon in my possession.'

On 24 February 1839 Gould arrived in Sydney. The town, a jumble of contradictions, did not impress him. At the time, Sydney was essentially a centre created around a convict population, yet eager to assert its status as the commercial and cultural capital of the continent. Grand public buildings of polished stone blossomed out of a mulch of mud huts and rough timber yards; chain-gangs laboured in the streets to clear tree-stumps and wagon ruts, while ladies paraded by beneath their parasols. It was a place stoked by both the puritan and the profligate, the missionaries and the military, the drunkards and the degenerates. Amid the cricket-clubs and colleges, crime was part of everyday life. Three years before, Darwin had arrived in Sydney proud to be an Englishman and left 'without sorrow or regret'. Sydney had some way to go before it was released from the burden of compulsive immigration, and before it could present to the world a face that was uniquely its own. While Gould's first impressions were not favourable, he was determined, as he wrote to Sir John Franklin, to keep an open mind: 'The heat and dust of Sydney is extremable neither does the presence of Drunkards which constantly present themselves in the streets add to the interest of the place, time and better acquaintance with the country will however perhaps enable me to speak better of it.'

Preparations for Gould's visit to his brother-in-law's station at Yarrundi had to be made, and it would be at least three days before Gould could escape the streets of Sydney to resume his 'favourite pursuits'. During this time he introduced himself to the Governor of New South Wales, who 'received me most kindly and offered me every assistance', and paid his respects to Captain Philip Parker King, who had commanded his expedition with Lady Franklin to Recherche Bay the previous December, and who was now resident in Sydney as Port Officer and Superintendent of Government Vessels. When, later in the year, Lady Franklin found herself in Sydney, she discussed with Captain King his impressions of their mutual friend, the ornithologist, and made a note in her diary concerning Gould's clumsy attempts at social etiquette: 'Captn. King agreed with me as to Mr Gould being an entirely uneducated man – he sd when he came here [to Sydney] he brought verbal introductions from Sir J. to difft. individuals & went up to them & sd. Sir J. Franklin desires me to give his compliments – he is in very good health & hopes you are the same.'

Whether Gould greeted all of his contacts with a similarly

gauche introduction or not, he made the best of his time in Sydney and called on, among others, the missionary Bishop of Australia, William Broughton, and Alexander MacLeay, first President of the Australian Museum at Sydney. It was in MacLeay's suburban garden that Gould saw his first living wattled talegalla or brush turkey roaming freely among the chickens. So extraordinary was this bird and so little was known about it, that previously naturalists had been divided as to whether it belonged to the vulture or the gallinaceous group. Gould was fascinated by its habit of collecting all the rubbish lying around in the shrubberies and borders and building a huge incubating mound in the middle of the lawn. Unfortunately, the creature drowned itself shortly after his visit by attempting to attack its own reflection in a bucket of water.

Gould also used his time to visit the great Australian explorer Charles Sturt at his home at Varroville outside Sydney. Sturt was just off to Adelaide with his family to take up the post of Surveyor-General of South Australia, and it was a matter of great good fortune that the two met at all. As a result, plans were laid for Gould to join Sturt's surveying expedition to the Murray River later that year. Sturt had already considerable experience of the countryside around Adelaide having in the previous year explored much of the land to the south of the Murray River, and crossed the Mount Lofty range on his overland expedition from Sydney to Adelaide. He would have been able to tell Gould about the wealth of unusual birds around Adelaide and persuade him of the necessity of collecting in the area. It was to be one of the most adventurous and rewarding expeditions Gould undertook during his stay in Australia.

Gould discovered in this hardy, middle-aged explorer, a man of great charm and ornithological ability. While visiting his house, Gould's ever-watchful eye was drawn to a large collection of watercolours painted by Sturt of parrots that he had collected on his various excursions. Gould offered an enormous sum for them on the spot. 'But', as Mrs Napier George Sturt described in her *Life of Charles Sturt*, 'these paintings had been the delight of Sturt's leisure; he was devoted to ornithology, and had collected the rarer specimens at great trouble and risk, and at no price would he part with the folio.' Gould's enthusiasm proved to be the kiss of death, for the collection disappeared without trace shortly after his visit. 'It was supposed', wrote Mrs Sturt, 'that Gould's remarks must have drawn the attention of some dishonest workman to the value of the drawings, for soon afterwards the military chest in which they were left, disappeared and was never again seen. Natives put upon the scent found military accoutrements and other articles thrown out of the same chest, so that the drawings were clearly the object of the theft. The fugitives were traced to Sydney.' The watercolours were never recovered.

The theft did not interfere with the good relations between the two men, however, and Gould, having secured a place on Sturt's forthcoming expedition to South Australia, set off towards the Liverpool Plains, north of Sydney. His expectations of the bird life were high, and he revelled in the sense of anticipation that gripped him before his departure. 'You, Sir,' he wrote to Sir John Franklin, 'fully understand this pleasure particularly when one's visits are directed to a fresh field abounding with novelties; if I find my labors as much rewarded here as they have been in V. D. Land, I shall consider myself amply paid.'

Gould's brother-in-law's station, 160 miles from Newcastle on the coast up the Hunter River at Yarrundi, far exceeded Gould's expectations both in size and in beauty. His homestead was situated on the Dart Brook, a tributary of the Hunter, nine miles from the great mountain chain known as the Liverpool Range. Since the days when he first settled at Yarrundi to farm sheep and cattle on the fertile Liverpool Plains, Stephen Coxen had accumulated about 12,000 acres of land and about as many head of livestock. More people had been drawn to the area since

Stephen had been there, but the settlement was still small and opportunities for the pioneer considerable.

Between 1838 and 1839, however, there arrived the first in a series of droughts. There were to be brief respites from the prolonged drought, one of which Gould was to be most fortunate to encounter, but the settlers were facing a period of unparalleled hardship. Many of them were bankrupted and forced to leave their farms. Eventually Stephen committed suicide in a Sydney hotel, leaving a note that stated, 'All the means I have in the world is on the table, 23/6d, and this I have borrowed'. Minutes before the poison he had taken took its final effect, a friend broke into the room to tell Stephen that a government maintenance had been granted him of two guineas a week. 'Had it come five minutes earlier,' he said when he heard the news, 'I might have been saved. I have taken poison.'

In 1839, though, Stephen was in good spirits. The drought had not yet taken a stranglehold, although the landscape looked parched and drier than it had for years. 'Stephen is doing well,' wrote Gould to Eliza when he arrived, 'better certainly than his neighbors. I find everything realized and more so than was stated. This is a pretty place far beyond what I anticipated. His *establishments* are very extensive and extremely well conducted. He has his failings however, I am sorry to say, the particulars of which I will explain when we meet. The drought is becoming very serious, no rain having fallen for nine months, in fact nothing but ruin presents itself. There is scarcely a blade of green grass in the colony. I could only have believed what I have seen. Stephen of course is suffering much but the distress is general.'

Gould had only a couple of weeks to spare here (he had been delayed a week in Maitland on the lower reaches of the Hunter while he waited to meet up with Stephen Coxen), having promised Eliza to return in time for the delivery of their child. His forays were therefore limited to the area around Yarrundi. He was joined on most of his hunting expeditions by two

Aborigines, Natty and Jemmy, 'two intelligent and faithful natives of the Yarrundi tribe', who became extremely attached to Gould and later accompanied him on his most ambitious and successful expedition to the Mokai and Namoi Rivers in December of that year. Eliza wrote to Mrs Mitchell at Broad Street about Gould's admiration for the Aborigines, and for Natty and Jemmy in particular:

'My dear Mrs Mitchell,
    Perceiving on a re-perusal of one of Mr Prince's letters that you are all feeling apprehensions of our being in danger from the natives, I hasten to relieve you from that alarm.
    I have also participated in your fears during John's absence at Sydney, but as I now believe quite needlessly. Indeed he considers a prudent man more safe with many of the black tribes than without their protection. There are, however, some hostile tribes far in the interior, but I trust he has sufficient prudence not to venture within their district. The tribes which inhabit the Upper Hunter and the adjacent parts of the colony are extremely harmless and well disposed. Two of them (the men, I mean) became quite fond of John during the short time he was with them, and expressed much regret at his leaving, begging him to come again and to bring his *gin*, meaning his wife, with him.'

Most of Gould's time on this first brief visit to Yarrundi was spent in pursuit of one particular bird – the *Menura* or lyre-bird that abounded in the cedar brush on the slopes of the Liverpool Range. It was a particularly elusive bird – the 'most shy and difficult to procure' – that Gould had ever come across. Its striking appearance and unusual habits, however, made it a prize worth concentrating every effort upon. It was a bird that would appeal to popular sentiment; indeed, Gould advocated its adoption as Australia's national bird, and later used it as a motif on the covers of his *Birds of Australia* series, showing the male lyre-bird with the beautiful lyre-shaped plumes of its tail raised in courtship.

Gould's determination to possess the lyre-bird took him over some of the roughest terrain around Yarrundi: into precipitous gullies and ravines, through dense woodland and tangled undergrowth, and through, as Gould put it, 'rugged, hot and suffocating brushes'. He later recalled his labours in the stifling, drought-stricken bush:

> 'Independently of climbing over rocks and fallen stumps of trees, the sportsman has to creep and crawl beneath and among the branches with the utmost caution, taking care only to advance when the bird's attention is occupied in singing, or in scratching up the leaves in search of food; to watch its actions it is necessary to remain perfectly motionless, not venturing to move even in the slightest degree, or it vanishes from sight as if by magic . . . None are so efficient in obtaining specimens as the naked black, whose noiseless and gliding steps enable him to steel upon it unheard and unperceived, and with a gun in his hand he rarely allows it to escape, and in many instances, he will even kill it with his own weapons.'

For Gould, accustomed to the European way of shooting birds, of which stealth is rarely a part, the task was not so easy. His somewhat unorthodox methods are explained in his letterpress for *Birds of Australia*: '[One] successful mode of procuring specimens is by wearing a tail of a full-plumaged male in the hat, keeping it constantly in motion, and concealing the person among the brushes, when the attention of the bird being arrested by the apparent intrusion of another of its own sex, it will be attracted within the range of the gun . . .'

When he was not ducking and weaving through the cedar brush with this curious hat on his head, Gould was sitting up at night on the Appletree Flats waiting to ambush a tawny-shouldered podargus or crouching next to a hole in the ground close to Stephen's garden gate waiting for the appearance of an unfamiliar little mouse. He also devoted a good deal of his investigations to the extraordinary satin bower-bird, a native of the area and so called for its remarkable habit of building and decorating a sort of stage-set, in which it would perform ritual mating dances. Gould had seen one of these bowers in the Sydney Museum, which had been presented by his brother-in-law Charles Coxen, and was determined to observe the construction in use in its place of origin:

'On visiting the cedar-brushes of the Liverpool range, I discovered several of these bowers or playing-places on the ground, under the shelter of the branches of overhanging trees, in the most retired part of the forest . . . The interest of this curious bower is much enhanced by the manner in which it is decorated with the most gaily-coloured articles that can be collected, such as the blue tail-feathers of the Rose-hill and Pennantian Parrakeets, bleached bones, the shells of snails, &c; some of the feathers are inserted among the twigs, while others with the bones and shells are strewn about near the entrances. The propensity of these birds to fly off with any attractive object, is so well known to the natives, that they always search the runs for any small missing article that may have been accidentally dropped in the bush. I myself found at the entrance of one of them a small neatly-worked tomahawk, of an inch and a half in length, together with some slips of blue cotton rags, which the birds had doubtless picked up at a deserted encampment of the natives.'

Although he was never able, despite his 'utmost endeavours', to find the actual nest and eggs of the satin bower-birds, the new light he was able to throw on their bower-building habits caused a considerable stir back in England. The fancy taste for ornaments and trinkets displayed by these peculiar birds appealed to the Victorian dilettante. Gould organised the shipment of several complete bowers back to London, where they were put on display at the Zoological Society. The name of John Gould soon became synonymous with the fantastic bower-bird, and gave a new ornithological slant to one of the popular songs of the day: 'Will you join in the evening and charm us as you ever have done at the piano', asked Gould of Mrs Owen shortly after his return, '. . . I shall be most glad to . . . sing "O Come to the Bower".'

Gould was thrown into a state of feverish excitement by his forays into the Australian bush, and, in a note written 'after dinner in a great hurry and with a bad pen', he told Eliza:

'I have so much to say that I cannot tell you half in a letter. You will be happy to hear that I am in excellent health and the colony agrees with me well. I have already made an expedition to the Liverpool Range. On the top of one of the highest peaks I mounted, and had a good view of the plains below.

The birds are all moulting consequently in a bad state being our excuse, so have killed many other things. We killed 9 Manura [sic] superba 3 of which I have in brine for dissection, 2 for skeletons, the rest in skin, a pair of which with several other fine birds I intend for Lady Franklin. Pray give my respects to their Excellencies and remember me kindly to all . . .'

Gould was clearly reluctant to return to Tasmania so soon, when so many novelties awaited him in and beyond the cedar-brushes of the Liverpool Range. Begrudgingly, he details his plans, as promised, to return to Hobart, emphasising, perhaps a little unfairly given his wife's condition, the inconvenience this would entail:

'I intend leaving this the first moment I can for Sydney hence to Hobart Town to see you, although it will verily interfere with my pursuits. But how I am to get conveyed with my luggage to Maitland neither Stephen or I can devise, three teams of bullocks having left Warra on their way from the plains to this place eight days ago and have not yet arrived. Warra is 35 miles distant. When we heard of them last they were cutting down trees for them to feed upon which is very generally adopted. As to water every river is dried up, and men are constantly employed in sucking wells. Cattle and sheep are dying in every direction. All this makes Stephen fretful and peevish as well it might. We shall contrive some way however to get me down to Maitland town and the end of the week, when I shall make my way to Hobart Town the best way I can.

I trust my dear Eliza you are still well and that you received my letter. You have my sincere hope and prayers for a safe delivery out of your troubles, and which I fear not will be granted to you, and if I am not with you at the time you will have something to present to me when I do.'

Despite the difficulties of his predicament however, Gould managed to return to Sydney in just over a fortnight after writing his letter to Eliza. Here, while he waited to embark the *Susannah Anne* bound for Hobart, he met with the officers from HMS *Beagle*, then on its third surveying voyage around the world under Commander John C. Wickham. With his admirable powers of persuasion, and perhaps the Darwin connection, Gould managed to enlist some of the crew: the Commander, Rear-Admiral Stokes, a Mr Dring, and, in particular, Dr Benjamin Bynoe, who promised to surrender to Gould any birds they might collect on the voyage for naming and describing. Ultimately, Bynoe was to provide some of the most magnificent specimens to be featured in the *Birds of Australia*, the silvery-crowned friar-bird, the painted finch and the yellow-rumped pardalote among them. (In a letter to the Zoological Society written on 10 May and read out at the scientific meeting in London on 8 October, Gould described 19 new species, 13 of which he stated he had received from Dr Bynoe of the *Beagle*.)

Gould was cutting it fine. He was in Bass's Straits on 16 April but he managed to be in Hobart for the birth of his son on 6 May. The 'little Tasmanian', a 'prodigious fellow' according to his mother, was named, appropriately, Franklin Tasman. If Eliza's mother had suffered anxieties over the birth of her fifth grandchild so far from home, her fears were allayed by a string of reassuring letters sent back to London. Even Gould wrote to the Broad Street household three days after the event. Eliza wrote to Mrs Mitchell on 30 May, 'We are, baby and I, still going on well and getting more strength every day. Never before have I had so good a time or taken so little medicine.' Three weeks later Eliza proudly elaborated, 'Baby, now three months old, is a very fine healthy child . . . He is the finest child about here and a great pet with all in the family.'

Despite the relief of a healthy baby and an uncomplicated birth, Eliza was still pitifully homesick for her mother, whom she feared might suffer 'one of her usual attacks' over the winter, and for her other children. 'How are the dear children?' she asked Mrs Mitchell in a letter of 28 May, 'Do they appear to retain any recollections of Papa and Mamma? I often please myself with anticipations of our return, picture to myself the joy and greetings when we meet, and I try to fancy the appearance of the dear children – they will be grown out of all knowledge.' The balls and parties given by the Franklins did little to buoy up her spirits, being, as she described herself, 'such a shy, reserved being', and she longed for the quiet, regular family life of home.

Only five days after the birth of his son, Gould was off again, this time to join Sturt in Adelaide for the expedition to the Murray River. He left Hobart on 11 May, and was in

Launceston by 18 May in time to board the *Black Joke*, together with 820 sheep, five horses and four other passengers, all bound for Adelaide. Gould's collection up to this time consisted, as the *Hobart Town Courier* related on 24 May, of 'about 800 specimens of birds, 70 of quadrupeds (several of which are new), more than 100 specimens preserved whole in spirits, and the nest[s] and eggs of above 70 species of birds, together with skeletons of all the principal forms. Mr Gould,' the piece continued, 'has left our shores for Southern [South Australia] and we cannot wish him better success than that which he has hitherto experienced.'

Gould's arrival on 1 June 1849 in the 'City of Adelaide' was decorously announced in the *South Australian*:

'Mr. Gould, the ornithologist, from London, is at present in Adelaide. Those who have seen his splendid drawings of the birds of other countries, will rejoice that those of our colony (many of which are exceedingly beautiful, and some of them new) will in turn become subjects for his pencil.'

If Gould had been disappointed by Sydney, he was amazed by Adelaide. The 'city' he saw – which had been founded less than three years earlier – was little more than 'a chaotic jumble of sheds and mud huts'. Here and there groups of immigrants, newly arrived from England, Ireland, or Germany, toiled on buildings or felled trees. Streets and squares had only just been marked out with pegs and twine amid the clutter. Among the remaining trees, which would one day distinguish themselves in avenues and parks, and over the heads of the labouring workmen rose clouds of parakeets and honey-eaters searching for food. In the countryside beyond Adelaide, tiny villages were just beginning to emerge, and acres of land were being cultivated for garden plots. As Gould wrote his impressions to Prince:

TRAGOPAN HASTINGSII.

⅓ Nat: Size.

Drawn from Nature and on Stone by E. Gould.   Printed by C. Hullmandel.

The western tragopan (*Tragopan hastingsii*), drawn from nature
on stone by Elizabeth Gould, from
A Century of Birds Hitherto Unfigured from the Himalaya Mountains.

EAGLE OWL.
Bubo maximus (*Flemm.*)

The eagle owl (*Bubo maximus*),
drawn on stone by Edward Lear,
from *The Birds of Europe*.

The Chinese bluepie (*Urocissa sinensis*)
by John Gould and H. C. Richter,
from *The Birds of Asia*.

The red bird of paradise (*Paradisea sanguinea*)
by William Hart, from
The Birds of New Guinea and the Adjacent Papuan Islands.

'I wish it were in my power to give you a faithful picture of this famed city of two years standing. People live in tents, and customs are so different from what they have been used to, that I really wonder how they reconcile themselves to their new mode of life. On the whole, however, I think South Australia may be considered as flourishing, and its condition will ultimately be prosperous.

The Zoological here, from what I have already seen, is likely to be of a most interesting description, totally different in its nature from that of Sydney, but probably approaching nearer in its character to the productions found beyond the Liverpool range, or what is more properly called the interior of New South Wales.'

Much of the time between Gould's arrival in Adelaide and the departure of Sturt's expedition was spent with Sturt and his family, or Governor Gawler, whose wife was a friend of Lear's, and in exploring the countryside around the city. In the space of just three weeks Gould visited the coast between Holdfast Bay and Port Adelaide, spent several days with James Hawker near the Horseshoe camp on the Onkaparinga, and shot his 'finest specimens', of which he was later to rename the Adelaide parakeet 'in the very streets of the city of Adelaide'. Blue-bellied, porphyry-crowned musk and little lorikeets also flocked around the town and provided Gould with easy targets. 'The incessant clamour kept up by multitudes of these birds baffles description', he noted, '. . . they feed together in perfect amity, and it is not unusual to see two or three species on the same branch. They are all so remarkably tame, that any number of shots may be fired among them without causing the slightest alarm to any but those that are actually wounded.'

On 17 June Gould set off with Sturt across the Mount Lofty range towards the Murray River. Although part of the same expedition, Gould's and Sturt's different objectives led them to work independently. Gould was provided 'through the kindness of Col. Gawler, the Governor, and Capt. Sturt, with 'horses, a cart, and a small company, with the view of reaching the Murray'.

Several commentators have claimed that on this expedition Gould's party was the first ever to reach the great western bend of the Murray overland from Adelaide. But we cannot be certain that Gould even got as far as the river at all. He himself says he 'spent five weeks entirely in the bush in the interior, partly on the ranges and partly on the belts of the Murray'. Although he had a magnificent view from the top of the Mount Lofty range of the Murray River, winding its course across the flats through a belt of dense dwarf eucalypti, there is no mention of his ever having reached its banks, let alone the remote western bend 100 miles away. How this rumour came about is uncertain. Gould might have been disorientated – his record of dates, distances and directions are sometimes unreliable; he might have exaggerated the achievements of his expedition – he would not have been the first explorer to do so; or he might have been misquoted by an over-enthusiastic press. Whatever the answer, the western bend of the Murray, some 100 miles north-east of Adelaide, was not reached by an overland route until October of that year, when at least two parties, including one from the survey department, are recorded as having made it.

Gould, it appears from his own references, was too busy collecting specimens to engage in an attempt to reach a remote part of the river. On 26 June he records shooting a wattle-cheeked honeyeater 'on the ranges near the Upper Torrens'; he 'killed a Bittern on the 1st July near the Murray, above Gleeson's Station'; collected red throats 'about forty miles north of Lake Alexandrina'; spotted a white-eyebrowed pomatorhinus 'near the bend of the river Murray' (perhaps it was this remark that led some to guess that it was the great west bend to which Gould referred); and shot a pink robin in 'a deep ravine under Mount Lofty' (this was the last time an example of this species, *Petroica rodinogaster* was seen in South Australia, although one or two reliable observers said they had seen specimens in the area many years previously). Apart from the time Gould would have spent stalking, shooting, skinning, and preparing these and a host of other birds, there would have been no possibility of him visiting all these different localities *and* reaching the west bend of the Murray. Not only were his carts and equipment a great encumbrance, but Gould was limited in the amount of ground he could cover by a scarcity of supplies. Sturt, who would have been able to take a much faster and direct route, only made it as far as the western approaches to the bend, and was so hard-pressed by lack of water that he had to bleed a horse to survive.

Gould's accounts imply that he spent much of the latter part of his expedition, because of a shortage of water, doubling backwards and forwards from the Mount Lofty range to the Murray Scrubs (later known as the Mallee), where most of the interesting birds were to be found:

'Having with difficulty crossed the range over an entire new country, and penetrated to the centre of the dense *Eucalypti* scrub alluded to, in which I spent a night and part of two days without water for my horses, I was compelled, much to my regret, to beat a hasty retreat back to the ranges, in the gullies of which I even found difficulty in obtaining water. During a week's stay under the ranges I made daily visits to this rich arboretum, which would have served me to investigate until this time without exhausting its treasures; but, alas! our provisions failing, we were obliged to re-trace our steps.'

Curtailed in his researches he may have been, but Gould still came away from the parched furnace of the scrubs with some of the rarest novelties yet of his collection and a vast number of specimens of every description. The black-backed superb warbler, the graceful ptilotis, red throat, white-fronted honey-eater, and chestnut-backed ground thrush all fell to his gun in these few exhausting but rewarding weeks. 'He has sent home another fine collection of upwards of 500 birds, 40 Quadrupeds etc. the result of 3 months labour in S. Australia', reported Prince to Jardine on 1 January 1840. 'His gratification was almost unbounded when on arriving at the Belts of the Murray he found not only numerous new Species but new forms of Birds. The Botany also of this part of Australia surpassed anything he had then seen. During the expedition which was undertaken in the depth of winter he passed 5 weeks in the bush without seeing a civilised being, sleeping occasionally in a Tent, sometimes under a cart but more frequently on the bare ground wrapped in a Kangaroo skin rug still he had not had a cold or the slightest disorder . . .' On 28 September 1839, Gould was to tell Jardine directly, 'I have visited South Australia, a part that has afforded me more novelties than any other I have visited.'

But the expedition also gave Gould a greater insight into the distribution of species over all Australia: it suggested to him, in comparison with his recent knowledge of New South Wales and Van Diemen's Land, which species might be stationary and which dispersed over a greater area. Sturt, with his unrivalled knowledge of Australian geography and zoology, had much to say on the subject, and his observations must have contributed greatly to Gould's views. In a letter to Dr Bennett in Sydney from the 'Upper part of the Torrens' on 23 June 1839, Gould remarked, 'Capt. Sturt has just joined me on our way to the Murray. . . . I am much pleased with my visit to this portion of Australia which will be attended by very important results regarding the Geographical range of species besides many new ones.' Gould was with the one man who could intro-duce him most convincingly to the alliances between geography and the native flora and fauna of the continent. But, despite his experience with Darwin, Gould failed to make any evolutionary observations from Sturt's information. He was more interested in pinpointing areas where he could find the most new species.

Sturt was convinced, for example, that a vast tract of fertile land, to rival even the prairies and bayous of America, lay undiscovered in central Australia. His evidence was the migration of the water hen (*Tribonyx ventralis*), which arrived in enormous numbers in November from somewhere in the north, and returned again after only two or three months. Later in the year, after the Murray Scrubs expedition, Gould was to find evidence to support this when he explored the Namoi River district in New South Wales. The sudden appearance of great flocks of harlequin bronzewings in the area led him to wonder, 'Whence has this fine bird made its appearance? . . . May we not reasonably suppose that it had migrated from the central region of this vast continent, which has yet much in store for future discovery? The great length of wing which this bird possesses, admirably adapts it for inhabiting such a country as the far interior is generally imagined to be . . .'

So confident was Sturt that a bountiful Garden of Eden existed somewhere in central Australia that in January 1843 he began to sue for funds for an expedition into the interior. Finance for the venture was, surprisingly perhaps, slow in coming. In May 1843 Gould was given the opportunity to repay Sturt for his generosity towards him while he was in Australia when Sturt appealed for help in generating support for the project in England. Sturt's own resources were low. Like many other colonials, the often devastating effects of the climate on local agriculture, and the inordinate expenses of living far from commercial centres had reduced him to impoverishment, but he was also finding little recompense from the government. 'I am a poorer man by some 2000£ than when I came to the Province', he told Gould, apologising for his inability to pay his subscription to *Birds of Australia*, 'and my salary has been reduced to the lowest figure and is far below what I enjoyed as a private Gentleman.'

Despite this obstacle, Sturt was determined to press ahead with the expedition, and on 19 May he appealed to Gould to exert any pressure he could on Lord Derby to encourage his son, Lord Stanley – who was then Secretary of State for the Colonies – to back the proposal. As Sturt told Gould:

'I wrote some six months ago to Lord Stanley, volunteering the general exploration of the Continent. I forwarded to him an estimate of the probable expense which I calculated at under £4000 with which sum I engaged to organize a Party to keep the field for two years. I stated my views to Lord Stanley as fully as I could, but I fear that the Exchequer is not in a condition to enable him to second my views, I think you will do me the credit to say that if any man can organize an Expedition for interior research I can, and that there are few if any here more qualified from experience for such a task. I had long intended making this proposition to the Secretary of State, but I was urged sooner to it perhaps because I felt that I was reduced in circumstances and thrown from my position in Society, and I hoped by a grand effort to establish a permanent claim on the consideration of Her Majesty's Government. I did not however communicate with any of my friends in England on the Subject save only Sir Ralph Darling. If however you hear anything of the matter, urge it as much as you can, and believe me I will leave no unconnected links in my track – and that in two years I will leave but little to be discovered here, if it please Providence to continue to me its wonted care. I hope you will not consider this is boasting for I am no boaster. A man has a right to take credit for past success and to attribute it to the real cause. There is no greater achievement in geographical Research remaining to be done than to lift the veil from before this unpenetrable and mysterious Continent. As I told you I think I have made observations which lead me to conclude that there are not only high but fertile lands on the verge of our intertropical Regions, and arguing from the Character of the Ranges here I should not be surprised to find in the more distant parts of the Continent, a Country vying with America in richness of Natural products, and I have very little doubts but that the discoveries to be made would amply repay the government any outlay it might make. I am writing hurriedly to you but I have thought it right to apprize you of this. . . .

. . . What name have you given to the Water Hen which migrates to us yearly – or which did migrate last year and the year before. It is dark green in plumage on the back with slate colour almost approaching to black below. It had been seen nine years before at Kangaroo island and in 1840 came to us in thousands – running about the streets and Gardens and into the houses as if it had never seen man before. Thinking they frequented the Creek I would not allow them to be shot but they bent down my grain and destroyed acres of it, and simultaneously disappeared some months after harvest – Where did they come from where do they go?

Pray let me hear from you and believe me

Ever most faithfully yours

Chas. Sturt

May 19th

But it was not until 22 April 1844 that Gould passed on Sturt's letter to Lord Derby with a cursory recommendation, and then only, it seems, after Lord Derby had mentioned the subject first. Gould, particularly where one of his most important and influential subscribers was concerned, felt exceedingly uncomfortable in the role of campaigner. As he wrote:

'My Lord, As you mentioned in one of your late letters

that Captn. Sturt was about to proceed upon a new expedition into the interior of Australia I beg to send for your perusal a Letter I have lately received from him and from which you will perceive that he has written to Lord Stanley; as I know no one better fitted for such a purpose than this enterprising and persevering Gentleman I do hope the Government may be disposed to second his views; perhaps, your Lordship, [sic] could obtain and favour me with some information on the subject; from the manner in which you referred to him in your Letter I am led to believe that some arrangement has already been made; pray say if such be the case.

At the end of Captn. Sturt's Letter your Lordship will find some interesting remarks upon a series of Gallinule to which I have given the name of *Tribonyx ventralis* and which I shall shortly publish in my work . . .'

It was late in 1844, a year of severe drought, that Sturt finally got official clearance for the expedition. He was to discover, after a staggering journey of 19 months and 3,450 miles, during which he came to within 150 miles of the centre of the continent, that no paradise such as he had imagined existed. Instead he and his men suffered terribly from scorching, thirst, and scurvy, from which one of his officers died. It was a miracle that any of them survived.

Sturt's Murray Scrubs expedition, restricted though it was in scale and goal in comparison with the great adventure six years later, also met with limited success. The parched and exhausted company eventually rendezvoused late in June or early in July in a homestead at Gawler, north of Adelaide. Mrs Mahoney, who lived there described years later:

'. . . an orderly arrived with dispatches for the Governor whom he was told to meet at our house. Next morning Colonel Gawler arrived with his A.D.C., Mr Gill . . . They had missed the track the night before so had to sleep in the bush with but one blanket each. They had come from the Murray. Evening brought the rest of the large party, and all dined with us – Captain and Mrs Sturt, Miss Gawler (about my age), and several gentlemen. Captain Sturt was the great explorer who had discovered the Murray River, and have been some years in New South Wales before coming to Adelaide, where he was Colonial Secretary. His brother Evelyn came, I remember, one morning soon after on his way to Adelaide, to send back provisions to his party, whom he had left quite out of flour. Gould, the ornithologist, was with him, and as they had been some days without anything to eat, except what they shot, they enjoyed a good breakfast. Gould gave me a lesson in skinning and stuffing; or rather preserving. His wonderful book shows the number of new birds (his wife painted them) he found in Australia, and when with us he was much pleased with a white kangaroo-skin he preserved, the first I had ever seen.'

Gould was perhaps the only member of the expedition fully satisfied with its outcome. He had reaped bountiful rewards where Sturt had again found his exploratory attempts frustrated. But time was pressing on, and Gould, who wanted to be back in New South Wales for the breeding season – he had assured Dr Bennett that he would 'be in Sydney not later than the middle of Augt' – returned directly to Adelaide.

After a brief visit to Kangaroo Island at the mouth of the Gulf of St Vincent, Gould set sail on 19 July on the *Katherine Stewart Forbes*, bound once again for Hobart. His expedition in the Murray Scrubs, then a 'wilderness, which the foot of white man had seldom trod and which no zoologist had ever explored', excited the imaginations of all who heard about it. Tales of his adventures preceded him.

Gould, however, was only an incidental explorer. Certainly

143

he had many of the qualities of a Sturt or a Leichhardt: he was energetic and determined, hardy and resilient. But he was not, unlike his colleague Captain Sturt, interested in geography for its own sake: his was not the thrill of covering distances, of making the first mark on the blank slate of terra incognita but the desire to discover new species, regardless of their place on the map. Geography played an important part for Gould in identifying the range of species, but he would have been happy to remain in the same spot if it would continue providing him with novelties. All the same, he undoubtedly encountered at least some 'new country' in his exhausting explorations in the scrubs; and he was rewarded by having three features in the area – Gould's Mount, Gould's Creek, and Gould's Range – named after him. His adventurousness, on a smaller scale than his friend Sturt's, and invested with considerable self-interest, met, in contrast, with unparalleled success.

CHAPTER 9

A Glorious Haul

FTER A FORTNIGHT in Hobart, where Eliza had been stationed for 11 months, Gould and his retinue packed everything, including Eliza's enumerable drawings, and left for Sydney. Much of what Gould had collected had been shipped back to London in sealed containers, where it was guarded like Pandora's box until release to the scientific world on his return. 'Mr G. has sent many specimens in spirits,' wrote Prince to Jardine, 'none of which have been opened as it is his wish that they shall not be shewn to anyone until his return.' Jardine's concern that this was not necessarily the best procedure, given that the specimens might be subject to decay, proved the case when several of the boxes, which Prince had been forced to break open, were found to contain moths and mildew. Even Gould's expert methods of preservation, vested as they were with such personal concern, were susceptible to the ravages of nature and a long sea voyage. Prince wrote to Gould describing the state of one consignment with something of a rebuke for the veteran taxidermist:

'The case on its arrival was too large to be got upstairs. I thereupon emptied it, tore it apart, carried it into the Drawing room, put it together again and restored the contents. How is it you put no camphor or turps into the cases: I have only opened two; one of these was the nest of *Cracticus hypoleneus* out of which several moths immediately flew . . .'

Another container was received in an even more chaotic condition:

'I yesterday got out the Box from the Marianne. I regret to say that in consequence of the contents having been packed before they were thoroughly dry they are all in a badly mildewed state, so much so that I shall have to dry each specimen individually and remove the mildew with a brush: the smallest of the two large skins has also lost the whole of the wool or hair from the head; The bottles too not having been properly secured, the spirit has escaped and caused the small boxes to fall to pieces forming an heterogenous mixture of nests, eggs broken and sound, tops, bottoms, and sides of boxes, plants, etc. The books are also much damaged in their bindings. . . . Every care shall be taken to prevent accidents from Fire Etc.'

Gould's deepest fears, however, were concerned with the dangers of dispatching his collection back to England unaccompanied. He passed on his concern in a letter to Prince, who was just as concerned at the receiving end. 'I regret you should have determined upon embarking any portion of your Property in V.D.L. because from the ignorance of the parties whose names you mention, I fear least they should turn out Sharks and it would be a source of deep mortification to me were you by these means to lose any portion of the advantages you have obtained by years of anxiety and deep mental study.'

Gould's fears were not exaggerated. Such was the demand

for rare and new species of any zoological description, that the colonies abounded with 'entrepreneurs' ready to steal a collection if they got the chance. The only hope for the naturalist was that the stolen specimens might re-surface in Europe, where he could then buy them back. Part of Gould's collection sent from Sydney shortly after his first visit to New South Wales only narrowly escaped being lost when it turned up in a warehouse that Gould happened to visit in Adelaide. The following year a consignment sent by his assistant Gilbert, who had remained behind in Australia, fell, much to Gould's dismay, into the wrong hands. As he wrote to Gilbert on 15 November 1840:

'You will be vexed to hear that some of the birds said to be in your last package from Sydney were missing: by what I can gather from your list they formed the whole of one of the smaller packages. This I much regret as it contained some of the most valuable things you had collected viz. the bird you call the Whiteheaded Swallow, all the Cinclosomas etc. These certainly were not in the box when it arrived in England and I fear were purloined from you either at Perth or Sydney, the tin box where in all probability you intended to put them not being opened at the Docks. When you write send me your opinion of this. If they should come to London they will probably get into the dealers' hands. I will look out for them. As you did at the Swan you will separate your collection sending one portion to me and retaining the other in case of accidents by sea. . . . But pray be careful what vessel you send them by. The Box sent by the Herald and which I accidentally found in a warehouse in South Australia was as near being lost as possible. Always write a letter saying to whom they are consigned in *Duplicate*.'

However great the risks of parting with one's collection, Gould could not have hoped to carry with him from Hobart the enormous quantities of skins, skeletons, eggs, nests, plants, and seeds he had amassed. He had no alternative but to trust his fortune to fate and the high seas. Perhaps for the first time he felt something akin to Eliza's feelings of separation from her children, as he watched his boxes and bottles winched aboard ship and slowly disappear over the horizon.

Gould was torn between the need to preserve his specimens and the desire to keep them a secret. Eventually reason prevailed, at least on the part of Prince, who requested of Jardine 'any information' he could offer for the preservation of Gould's precious collection. 'My Thanks are due,' Prince wrote to Jardine on 7 April 1840, 'for your instructions respecting the specimens in spirits, which I have since carefully examined, separated and supplied with fresh spirit; many of them I regret to add are much decomposed and I fear some may at last prove useless.'

The voyage from Van Diemen's Land to Sydney on the *Mary Ann*, with his wife, children, servants, and the remains of his baggage proved to be a 'tedious and uncomfortable' one; different by far from Gould's first, joyous passage to New South Wales six months earlier. He had, moreover, the trials of business to look forward to before he could get back to work.

In Sydney, while they waited for a steamer to take them up the coast to Newcastle, the Goulds stayed with Dr George Bennett, secretary to the Australian Museum. Gould spent his time in the city preparing for the final and most ambitious expedition of all his trips in Australia. 'Nearly a fortnight', he complained, was spent in Sydney organising his journey to the other side of the Liverpool Range – that remote and tantalising region that had eluded him on his first visit to Yarrundi because he had had to return to his wife in Hobart. This time he would take Eliza with him all the way to her brother's homestead at Yarrundi, so he could be sure of being in the area for at least three months. Eliza was under no illusions that this meant they would be spending more time together: 'John, of course, will not remain there long,' she told her mother, 'but be wandering off in some direction . . .'.

Gould still intended to do a lot of collecting on the Yarrundi-side of the Liverpool Range, but the main focus of his efforts was to be across the mountains, on the remote plains around the Namoi and Mokai rivers, an area of the interior that few settlers had yet penetrated. Apart from the forays of his brother-in-law, Charles Coxen, this area had never been explored for birds. Gould projected the expedition would take about two months. He was assigned 'two or three trusty convict servants' by Sir George Gipps, Governor-in-Chief of New South Wales, together with government 'tents, and the necessary utensils and materials for leading the life of a bushman'.

Having settled all the necessary arrangements, the couple at last set off from Sydney on 14 September for Newcastle, and the mouth of the Hunter River. It was Gould's thirty-fifth birthday, although no mention is made of it in Eliza's journal. Gould wrote, though, to Prince, and gave him a few rare but welcome words of recognition for all his trouble. Touchingly, Prince acknowledged the letter almost at once:

No. 25

> London, 20 Broad Street
> Golden Square. Feb 5: 1840

Dear Sir,

Your very welcome Letters of the 20th of Aug. and 14th of Septr. reached us on the 23rd and 25th of Jan: they were a joyful relief to us all and were the most acceptable to me since for the first time you acknowledge I have been tried and not found wanting: believe me it will always be my highest gratification to merit the good opinions of every one but of none more than yourself: and the more confidence you repose in me the more strenuous will be my efforts to prove myself worthy of it.

I am delighted to find you have been so successful in

your Expedition: I heartily pray you may continue so to
the end and safely return to Old England: I am much
pleased too at hearing your health has not been affected in
the slightest degree: pray be careful and never forget that
upon your safety, hinges the welfare of numbers.'

On arrival, after an uncomfortable night tossing about on board
the steamer, Mrs Gould, her maid Mary, and the baby were
lodged at an inn in Newcastle, while Gould took the opportunity of visiting some of the islands in the estuary. He soon
returned, however, to collect Eliza, who, despite her misgivings
about leaving the baby who had been poorly since their stay in
Sydney, accompanied her husband for the day to sketch the
living bird and plant specimens on Mosquito Island. In an
extract from a journal (the only one to survive from the trip),
which Eliza kept during that September, she describes this
brief outing at the command of her restless husband:

'Monday 16th. John went over to Mosquito Island where
he has established James for a few days with a government
tent, man and fitting-up. Came home in the evening he
says purposely to take me over next morn early.
   17th. Rose at six dressed hastily got some breakfast and
had a pleasant row to the island. found the tent pitched in
a cleard [sic] spot in the midst of the bush where nature
appeared in her wild luxuriance. The Immense parasites
twining round the trees taking root some of them at the
tops of the trees and hanging down to the ground, others
surrounding the trees like a crown – heard the bell bird
with his incessant ting ting, the coachwhip bird &c. – a
heavy shower of rain accompanied by lightning – soon
cleared up – every green thing looked more beautiful for its
sprinkling.'

The forests on these islands abounded in honey-eaters, regent-birds, satin-birds and wood-pigeons; Mosquito Island was also where Gould saw his first live brush turkey, or wattled talegalla, scratching about on the forest floor. It also yielded the rare *Sphecotheres maxillaris* or southern sphecotheres, and a fine specimen of the southern fig bird (*Sphecotheres vieillori*) among others; and provided Mrs Gould with plenty of ideas for botanical backgrounds. But it was only a momentary diversion, and the couple hastened up river in the steamer, *William IV*, to the town of Maitland to await the arrival of the bullock carts from Yarrundi. Here, in the Rose & Thistle Inn, in this prosperous trading-post that was the last port-of-call for traffic from Sydney, Mrs Gould's journal reports a return to the familiar routine: 'employed all day making drawings' (24 September); 'drawing all day'; 'drew all day'; while her labours were relieved by the occasional walk in the cool air of the early morning or evening:

'Walked on the race course before breakfast the air balmy and very delightful, great numbers of the blue mountain parrots were making their morning meal on a large kind of the Eucalypti – two of the beautiful Nankeen night herons passed over our heads and we heard the curious note of the coul [cowl] bird or bald-headed friar – returned with an excellent appetite – drew all day – in the evening John called me to look at the skin of a snake more than six feet long which James shot in the act of ascending a tree – also brought me some beautiful specimens of a climbing plant bearing thick clusters of cream colour blossoms.'

Gould recorded the fortnight waiting for the arrival of Stephen Coxen's bullock train in Maitland as being 'pleasantly spent in the bushes of the Hunter among the Satin [,] Regent [, &] Cat Birds.' He also took the opportunity of writing, for the first time, to Sir William Jardine:

151

Maitland River Hunter
New South Wales
Sept. 28th 1839

Dear Sir William:

You have doubtless heard from time to time some little of my movements on this side of the globe and if I have not written more fully and frequently to my friends in England I trust this will be attributed to the very pressing engagements which have fully occupied my time and mind.

You may readily imagine the extreme gratification I feel in visiting this fine country, teeming as it does with so many interesting and beautiful productions. My success up to the present time has been greater than I could have anticipated both as regards obtaining much information that is entirely new as well as in bringing together one of the finest collections that has ever been formed. I have as a matter of course made a point of attending to those particulars which hereto for [sic] have been overlooked not only by collecting the birds in their various changes of plumage but by preserving all the principal forms for dissection as well as preparing skeletons of the same . . .

Mrs Gould who I am happy to say is quite well has been fully occupied in making drawings of the soft parts of the birds together with appropriate plants flowers berries etc. which will be introduced into my work and I trust from the fund of information I shall be enabled to add that the book will not be void of interest especially to all lovers of ornithology . . .

This is one of the finest climates in the world. Neither myself or anyone of my party having received an hours illness since our arriving though constantly exposed night and day to the bush in all kinds of weather. As regards myself I rarely ever tire or find the day too long though I am constantly walking a circumstance which being considered much to my health being better able to bear fatigue than when last I walked over the hills with you and I found your advice not to take spirits very very judicious. I find tea to be the best and most satisfying bush drink that can be taken . . .'

The drays and bullocks Stephen had sent from Yarrundi eventually arrived on 27 September, the 'tent was struck' where Gould's men had encamped outside town with all the provisions, and was loaded up on the carts for the long, slow haul back to the reaches of the Upper Hunter. Gould was to follow two days later. He was longing, as he told Jardine, to be back among his beloved lyre-birds, by now in the last stages of their extraordinary seasonal plumage. 'I am sure you would be delighted to spend a week among the *Menuras*' Gould wrote, 'as I hope to do almost immediately. It being my intention to encamp near their haunts in order if possible to obtain their eggs and learn something of their habits of nidification . . . It is a cheerful active bird singing and mocking all the birds of the forest . . .' Gould would also be reunited with Natty and Jemmy, whom he planned to take on the Namoi expedition. He described to Jardine how indispensable the Aborigines were:

'I find the natives very useful in assisting being rarely ever without a tribe or a portion of a tribe with me when in their neighbourhood. They are nearly all excellent and dead shots and are especially fond of shooting. I frequently give into their hands my best guns and never found them in the slightest degree disposed to take advantage. I am of course not speaking of those from the interior it is necessary to be more guarded with them particularly those of the Namoi and Juden [?] parts, which I am now about to visit. I shall then in fact have to stay on the alert.'

On Monday 29 September the Goulds left Maitland by cart for Yarrundi. News had reached them while they were still in

Sydney that the drought that had afflicted the Liverpool Plains earlier in the year had finally broken with abundant rains. Land prices consequently had risen, and Stephen was suddenly a rich man. But nothing could have prepared Gould for the transformation he witnessed on the journey back to Yarrundi from Maitland. Of the drought he had experienced on his last visit, Gould wrote that, 'It is easier for the imagination to conceive than the pen to depict the horrors of so dreadful a visitation.' Now, following the rains, the wild Australian bush was as meek and hospitable as the English countryside, with only a casual reminder of its former ferocity. 'This part of the country', wrote Eliza, 'resembles for miles a succession of parks thickly dotted with tall slender trees of the Eucalypti kind and the ground covered with luxuriant verdure where during the past season not a blade of grass was to be seen the bones of many bullocks which died on their way from the upper Hunter to Maitland rill still remain by the roadside a momento [sic] of the excessive drought.'

Gould and his party arrived at Stephen Coxen's station the day after a night spent on Patrick's Plains. Gould found he 'had arrived at a good time, the birds having just commenced breeding', and was immediately off to resume his researches with Natty and Jemmy in the cedar brushes of the Liverpool Range and on the nearby stretches of the Dart Brook. Eliza settled in for a four-month stint of painting and drawing.

During this time Gould found more new species: the red-chested quail (*Turnix pyrrhothorax*), the little eagle (*Hieraaetus morphnoides*), the black-breasted buzzard (*Hamirostra melanosterna*) and the red-browed tree-creeper (*Climacteris erythrops*) among them. And, thanks to Natty's agile tree-climbing, Gould secured the eggs of numerous birds, including the rare whistling eagle and the spotted-sided finch. A month later, just before he was about to set off on his journey 'into the interior by way of Namoi', Gould wrote to Sir John Franklin in Hobart of his progress so far: 'After spending a fortnight in the lowland brushes I proceeded to the upper districts and the Liverpool ranges whence I have just returned having made several discoveries of new species both of Birds and Quadrupeds, of the latter I believe I have two new kinds of Kangaroo.'

The elusive lyre-bird, though, despite all Gould's efforts, was living up to its reputation. 'I have not yet succeeded in finding the breeding haunts of the *Manura* [sic],' he confessed to Franklin. Gould never found the nests of the lyre-birds. Years later, on Gilbert's second visit to the colony, he asked specifically that Gilbert should investigate the 'nidification of the Menura'. With regard to the rest of the local ornithology around Yarrundi he was more confident, and in his instructions to Gilbert written on 20 January 1824, he stipulated, 'The Brushes are not worth hunting for the birds.'

At last, some time in mid-November, Gould was ready to embark upon his Namoi expedition. 'John is gone across the Liverpool Plains to the Namoi', wrote Eliza to her mother on 6 December from Yarrundi, 'he expects to find much to interest him there. It is particularly unpleasant to us to be so frequently separated, but of course my going with him is out of the question; he will sleep under a tent all the time.' The party comprised five Europeans (Gould, his servant James Benstead, and the three 'prisoners of the Crown'), and, naturally, Natty and Jemmy. It was to prove the most rewarding of any of Gould's explorations in Australia.

The abundant rains that came in the weeks previous to Gould's arrival had attracted birds in their thousands, some of which had never been seen in the area before. The area was crawling with caterpillars, upon which vast flocks of birds – including the beautiful straw-necked ibises – descended, and, in their wake, hundreds of hawks of three or four different species hovered, attracted by the sudden abundance of prey. Clouds of little-crested parrots and rose-breasted cockatoos swarmed upon the woods that were dotted here and there over the grasslands.

The country beyond the mountain range was completely different from the bushland that bordered the coast and to which Gould had so far been confined. Here were great plains where grasses of completely different species grew; the woodlands were not tangled, impenetrable forests but sparse copses. Emus wandered over the open country hooting their hollow cries; and troops of kangaroos, some of them of gigantic size, grazed peacefully in their primitive pastures. The hinterland of the Liverpool Range in the summer of 1839 was a resplendent, if temporary, Eden. Sheep-shearers and stock-keepers had also descended on the plains to take advantage of the excellent grazing, and provided Gould with company from time to time on his journey. It was a time of plenty for everyone; for Gould it was a period of unparalleled excitement and surprises.

By December Gould had crossed the Liverpool Range and set up camp on the banks of the Mokai. The first momentous occasion for the ornithologist occurred as he strolled beside the river at sunrise and caught sight of a 'new and beautiful pigeon', a bird never before encountered by any of the natives he questioned or any of the stockmen at the outstations. The bird, which Gould called the harlequin bronzewing, and which later became known as the flock pigeon, rose from the water beside him and alighted on the ground 40 yards away, just within reach of Gould's expert aim. He expected it to be the only example of its species on which he would set eyes.

A fortnight later, on the plains of the Namoi, Gould was thrilled to see the same bird in such great numbers that, when Natty drove them up before him, 'eight fell to a single discharge' of his gun. The coincidence of the bronzewings' appearance and Gould's arrival did not escape comment by the local Aborigines, who had, like Natty and Jemmy, attached themselves to the strange English bird-man. 'Both the settlers and the natives assured me', wrote Gould, 'that they [the bronzewings] had suddenly arrived, and that they had never been seen in that part of the country. The Aborigines who were with me, and of whom I must speak in the highest praise, for the readiness with which they rendered me their assistance, affirmed, upon learning the nature of my pursuits, that they had come to meet me.'

A week later, Gould described the moment when, as he and his companions were returning from hunting kangaroos on a distant part of the Namoi plain, they approached a small group of acacia trees, and he met with the bronzewing once again:

'Natty suddenly called out, "Look massa"; in an instant the air before us seemed literally filled with a dense mass of these birds, which had suddenly rose from under the trees at his exclamation; we had scarcely time to raise our guns before they were seventy or eighty yards off; our united discharge, however, brought down eight additional specimens, all of which being merely winged and flutter- ing about, attracted the attention of our kangaroo dogs, and it was with the greatest difficulty that they could be prevented from tearing them to pieces; in the midst of the scramble, a kite, with the utmost audacity, came to the attack, and would doubtless, in spite of our presence, have carried off his share, had not the contents of my second barrel stopped his career. This was the last time I met with the Harlequin Bronzewing.'

Another strange bird had descended in great flocks on the Mokai, much to Gould's delight as only a single specimen had ever been delivered to England prior to 1839, which was in the collection of the Linnean Society. The warbling grass parakeet or *Melopsittacus undulatus* was a familiar bird to the Aborigines, who called it the 'betcherrygah'. But they had never before known it to come near the Namoi. There was, once more, only one explanation. 'The Black fellows of the Upper Hunter told me', wrote Gould in a letter to E. P. Ramsay in Sydney in 1866, 'that the little Melopsittacus undulatus had come to meet me, for they had never seen the bird in that district until the year I arrived.' That December the birds thronged near Gould's encampment near Brezi. They were breeding in huge numbers in all the hollows of the tall eucalypti that grew on the banks of the Mokai, and fed in flocks of many hundreds upon the grass seeds on the plains, descending in great noisy crowds on the pools by Gould's tents to drink. In the intense heat of the day most of them fell silent and withdrew into the gum trees, where they disappeared as if by magic among the bright green leaves.

Gould was charmed by these little parrots whose 'extreme cheerfulness of disposition and sprightliness of manner' he said 'render it an especial favourite with all who have had an opportunity of seeing it alive'. How great an attraction, he accurately predicted, would the betcherrygah prove in the parlours and drawing-rooms of the great British gentry, if he could just get it home alive. 'The beauty and interesting nature of this little bird', Gould wrote, 'naturally made me anxious to bring home living examples; I accordingly captured about twenty fully fledged birds, and kept them alive for some time; but the difficulties necessarily attendant upon travelling in a new country rendering it impracticable to afford them the attention they required, I regret to say the whole were lost.'

Charles Coxen, however, had managed successfully to raise some of the birds at his homestead on the Peel, and gave Gould four on his return from the Namoi. Two of these survived the arduous voyage round Cape Horn in the middle of winter and arrived to a grand reception in London. The tiny parrot was a thrilling novelty, and before long every visitor to the Australian colonies was attempting to catch them and bring them back as souvenirs. As Gould described to his Victorian reader, as yet unacquainted with this most common and popular of pets:

'As cage birds they are as interesting as can possibly be imagined; for, independently of their highly ornamental appearance, they differ from all the other members of their family that I am acquainted with, in having a most animated and pleasing song; besides which, they are constantly billing, cooing, and feeding each other, and assuming every possible variety of graceful position. Their inward warbling song, which cannot be described, is unceasingly poured forth from noon to night, and is even continued throughout the night if they are placed in a room with lights, and where an animated conversation is carried on.'

One can imagine the crescendo of noise in Gould's sitting-room in Broad Street as crowds of visitors, exclaiming with delight, were greeted with the excitable chatterings of the little betcherrygahs. The birds added to Gould's party repertoire. Professor Richard Owen's (the eminent anatomist) wife noted in her diary for 27 March 1841: 'Lord Northampton's evening party. Richard very tired, and thought he would not go . . . after some dinner R felt better . . . He was glad afterwards he went, for Prince Albert was there, and Mr. Gould brought his pretty singing New South Wales parrots.' Gould would surely have been amazed to know quite how commonplace his precious little parrots, or 'budgerigars' as they were called eventually, would become.

Gould and his party swept down the Mokai River to its junction with the Peel River, ransacking the plains for novelties on either side. On 11 December he shot the first-ever specimen of the crimson chat as he traversed the forest bordering the River Peel, to the east of the Liverpool Plains. 'As may be supposed', he wrote, 'the sight of a bird of such beauty, which, moreover, was entirely new to me, excited so strong a desire to possess it that scarcely a moment elapsed before it was dead and in my hand.' He continued down the Namoi for 150 miles towards the Brigalow Scrubs – a distance of 400 miles from the coastline, and the farthest he went into the interior – slaughtering en route a plethora of birds, many of which – the white-backed swallow, the black-eared cuckoo, the yellow-throated miner, the striated grass-wren, the red-backed kingfisher and another bower-bird of the spotted variety to name but a few – were completely new to science.

In these heady days, Gould's acquisitiveness seems distinctly unsportsmanlike. 'Of its mode of nidification', Gould wrote of the nightjar he found on the lower Namoi, 'I can speak with confidence having seen many pairs breeding during my rambles in the woods . . . In every instance one of the birds was sitting on the eggs and the other perched on a neighbouring bough, both invariably asleep. That the male participates in the duty of incubation I ascertained by having shot a bird on the nest which on dissection proved to be a male.'

Neither did the square-tailed kite, which Gould found near Scone, receive mercy at the end of Gould's barrel. 'I met with it', Gould recalled, 'in various parts of New South Wales, and on the plains of the interior, still it is by no means abundant, and persons who had long been resident in the colony knew little about it. I had, however, the good fortune not only to kill the bird myself, but, in one instance, to find its nest, from which I shot a female.'

Gould was in seventh heaven. As well as finding all about him an array of ornithological novelties of unparalleled beauty and originality, he had discovered a region abounding in extraordinary marsupials: giant kangaroos that could outrun a dog, duck-billed platypi, opossums, koalas, and spiny ant-eaters. Not satisfied with the haul of birds that he would bring back to Britain, Gould began to consider collecting marsupials as well. Here was another subject, he suddenly realised, that was as yet unpublished; a perfect companion work to his proposed *Birds of Australia*. 'If the Birds of Australia had not received that degree of attention from the scientific ornithologist which their interest demanded', he wrote in his preface to *Mammals of Australia*, 'I can assert, without fear of contradiction, that its highly curious and interesting Mammals have been still less investigated. It was not however, until I arrived in the country, and found myself surrounded by objects as strange as if I had been transported to another planet, that I conceived the idea of devoting a portion of my attention to the mammalian class of its extraordinary fauna.'

Gould could not resist plucking every fruit that Australia dangled before him. In that weird and wonderful land of the interior, under the inspirational guidance of Natty and Jemmy, 'these children of nature' as he called them, the mammals too began to captivate the ornithologist's imagination. For the first time he was tempted to commit scientific adultery. In his preface he elaborates, almost apologetically, on this process of seduction:

'Tired by a long and laborious day's walk under a burning sun, I frequently encamped for the night by the side of a river, a natural pond, or a water-hole, and before retiring to rest not unfrequently stretched my weary body on the river's bank; while thus reposing, the surface of the water was often disturbed by the little concentric circles formed by the *Ornithorhynchus* [duck-billed platypus], or perhaps an Echnidna [anteater] came trotting up towards me. With such scenes as these continually around me, is it surprising that I should have entertained the idea of collecting examples of the indigenous Mammals of a country whose ornithological productions I had gone out expressly to investigate? . . .'

Skins and skeletons of mammals, from the giant kangaroo to the tiny kangaroo rat, were hauled on board the groaning bullock carts to join the piles of feathered carcases already stacked away as part of Gould's mammoth collection. The expedition had been rewarded with trophies beyond his greatest dreams, and he returned to Yarrundi, sometime in mid-January, laden down with booty.

Gould had escaped just in time: a week after his departure a deluge of rain hit the Namoi plains, and the flash-flood that followed wrought havoc on the countryside. Tragedy struck two settlers, a Lieutenant Lowe and his nephew, who were among those grazing their sheep on the Namoi plains. Gould had stayed with them in their hut on his way to and from Namoi. In an extract from the introduction to the *Birds of Australia*, Gould describes the fate that befell these two men who lived and died at the mercy of the contrary Australian climate:

'Seven days after my departure from their dwelling heavy rains suddenly set in; the mountain-streams swelled into foaming torrents, filling the deep gullies; the rivers rose, some to the height of forty feet, bearing all before them. The Namoi having widely overflowed its banks, rolled along with impetuous fury, sweeping away the huts of the stock-keepers in its course, tearing up trees, and hurrying affrighted men and flocks to destruction. Before there was time to escape, the hut in which Liet. Lowe and his nephew were sojourning was torn up and washed away, and the nephew and two men, overwhelmed by the torrent, sank and perished. Lieut. Lowe stripped to swim, and getting on the trunk of an uprooted tree, hoped to be carried down the eddying flood to some part where he could obtain assistance. But he floated into the midst of a sea of water stretching as far as he could discern on every side around him. Here he slowly drifted; the rains had ceased, the thermometer was at 100, a glaring sun and a coppery sky were above him; he looked in vain for help, but no prospect of escape animated him, and the hot sun began its dreadful work. His skin blistered, dried, became parched and hard, like the bark of a tree, and life began to ebb. At length assistance arrived – it came too late; he was indeed just alive, but died almost immediately. He was scorched to death.'

Yarrundi, on the other side of the Liverpool Range, was not badly affected by the rains. Gould's wife had spent the months that he was away over Christmas and New Year patiently labouring over her sketchbooks and specimens, and had a stack of botanical and ornithological drawings to present him with when he returned.

The region finally being exhausted to Gould's satisfaction, the couple returned to Sydney. They had been in the outback for four months, and during that time it had become clear to Gould that his projected trips to New Zealand, Moreton Bay, and Norfolk Island were unnecessary. Later Gould was to cross New Zealand off his list altogether. In a letter to Gilbert dated 16 November, 1840 he wrote: 'As I find the fauna of New Holland quite distinct and that of New Zealand to belong to an entirely different group of Islands I have determined not to include it in the present work; you will therefore *not go there* . . .'

He had been successful 'beyond his expectations', and was now in possession of the most 'glorious haul' yet to be made by an ornithologist in the field. He could not wait to see it all back safely in England and to start on his next publication. 'I am happy to say,' Gould wrote to Sir John Franklin, with uncharacteristic modesty and a nod to the Creator, 'that my last trip to the interior has been productive of much that is interesting, having discovered many novelties both in birds and quadrupeds, my whole journeys in fact to these colonies have been most auspicious ones and I return satisfied and especially thankful for what our almighty providence has in his infinite goodness allowed me to see. My next hope is to be spared to publish the results of my enquiries.'

To cover the areas of special interest that Gould had been unable to investigate personally, he had to rely on a network of collectors he had recruited in the course of his travels. He left behind him in Australia an army of naturalists, both amateur and professional, to whom were assigned a multitude of tasks to increase and perfect his collection. The house in Broad Street was to be inundated in the years to come with hopeful contributions from naval captains, clergymen, convicts, sheep-farmers, and soldiers, as well as Gould's own specially appointed collectors. Port Essington, the most remote, dangerous, and perhaps most promising, hunting ground left on his itinerary, and which Gould himself had intended to explore before his triumphant visit to Namoi, was designated to his assistant Gilbert, who was staying an extra year in Australia.

Gould had arranged to rendezvous with Gilbert in Sydney before leaving for England, but so eager was he to set sail with his precious cargo that when it seemed Gilbert might have been delayed in Western Australia, he left without seeing him. Three weeks after the Goulds had sailed in the *Kinnear*, Gilbert arrived in Sydney, exhausted, penniless, and with a glorious haul of his own – only to be met by a letter of further instructions.

The four-month voyage back to england via the South Pacific and Cape Horn was spent much in the same way as the outward voyage. Time and again Gould was lowered over the side of the ship in a little dinghy at the command of the accommodating Captain – called, appropriately, Mallard – where he floated happily for hours among his birds. Few new species were added to his oceanic collection, but Gould was delighted to spend his time recording again in minute detail his encounters with familiar friends.

Gould arrived home on 18 August proclaiming the triumph of his endeavours to the scientific world. In his own words he returned with a 'rich harvest of knowledge', 'the finest collection extant'; in the words of Prince, 'His collections of Birds, quadrupeds, nests and eggs are unquestionably the finest that have ever been brought home'. In a fit of feverish activity Gould proceeded to publish a series of lectures and papers as quickly as he could possibly produce them; on 25 August he exhibited and named at the Zoological Society six new species of kangaroo and presented his information on the bowerbirds; on 8 September he read a paper at a scientific meeting on the extraordinary brush turkey, or *Alectura*, among a selection of others; on 13 October he gave an account of the *Ocellated leipoa*; on 10 November he exhibited at the Society 50 of his new Australian birds, which he characterised in subsequent meetings; and so on, until he had presented all his new trophies from both the bird and animal kingdoms to the awestruck Zoological members.

At the same time Gould began to organise the production of the plates and letterpress for *Birds of Australia*. Indeed, 'so much engaged with his Collections and twenty other matters' was he that he left it to Prince to contact his colleagues and subscribers to inform them of the 'results of his journey'. When he finally got around to writing to Jardine on 9 September, it was with only a cursory description of his achievements on the other side of the world. 'My collections, I am happy to say' he wrote, despite his own and Prince's fears, 'have all arrived in safety and I can now scarcely tell how so large a mass was got together in so short a time.' As a result, he continued, 'It is quite impossible for me in a letter to give you a detailed account of my proceedings my time being at present fully occupied in unpacking arrangements and other matters.' So little time had he to spare, indeed, that he declined Jardine's invitation to shoot with him in Scotland, an opportunity he had in the past seldom passed over. 'Although, God knows', he said, 'the poor Birds ought to have a holiday from me for years to come, for I have been incessantly shooting and my own gun was not the only one that waged war upon them. I have scarcely even been without natives accompanying me who also shoot with facility.'

When he wrote to the Prince of Canino five months later he had a little more time to describe the achievements of his two years 'in the wilds of Australia':

'The interval spent from my native shores were some of the happiest days of my life . . . The results of my journey cannot, I think, but be attended with great advantage to science returning as I have loaded with novelties in every department of zoology; besides much entirely [new] information respecting the habits and economy of the singular forms peculiar to Australia. Independently of a great number of new Birds I succeeded in procuring the nests and eggs of at least two thirds of the species inhabiting that interesting region. In Mammalia I have also done much having nearly doubled the before known numbers of species of Kangaroos many of them large and noble animals. I also procured on my voyage out and home a magnificent Collection of the Procellaridae and noted with accuracy the different Lati[tude] and Long. in which I observed them.

Since my return I have been busily employed in *recommencing* my work on the Birds of Australia . . . If you can procur [sic] me *any supporters* for this, the finest and most perfect work I have yet attempted, you will be doing me a real kindness; you are aware that neither pains or expense will be spared to render it complete.'

Mrs Gould, delighted to be home again and reunited with her children, set to work once more on the lithographic stones at her desk. Part 1 of *Birds of Australia*, containing 17 plates, was scheduled for publication on 1 December, less than four-and-a-half months after the Gould's return to England. The total cost of the expedition had come to roughly £2,000 – half of which Gould was to recoup less than seven years later when he sold his collection of Australian birds for £1,000. The publication of *Birds of Australia* proved ample financial reward for his initial outlay. Figuring a total of 600 plates in seven volumes (of which 328 species were new to science and named by Gould), it was his most expensive production to date. At £115 (today a first edition set of *Birds of Australia* is worth £100,000-£150,000) 283 subscribers signed up for the privilege of owning a copy. It was to take eight years to produce the entire series.

Some scientists were dissatisfied with Gould's decision to target such an elitist market. Swainson was suggesting, even before Gould had returned to England, that he would do better to make his books more generally affordable. Gould, he said in his book on *Taxidermy*, was 'a zealous and very able ornithologist, now travelling in Australia, who has published some valuable, although very expensive works upon birds. . . . We trust the author will hereafter reprint these expensive volumes in such a form as that they may be accessible to naturalists; and thereby diffuse science, instead of restricting it to those only who are wealthy.' Eventually Gould was to take Swainson's point, and in 1865 published a *Handbook to the Birds of Australia* in two more manageable royal octavo volumes, with no illustrations and a revised text.

Gould's *Birds of Australia* remains the most comprehensive work ever produced on the subject. Comparatively few species have been added to Gould's list of native Australian birds; at least two, the Lord Howe white-eye (*Zosterops strenua*) and the Norfolk Island parrot (*Nestor productus*) have become now extinct, and at least 20 species are on the verge of extinction or extremely rare. Gould's vast, lavish, hand-coloured volumes, with copious descriptions of every bird, provide an important historical record and a database for research, even today. In the eyes of Australia, Gould has become the 'Father of Ornithology'.

Gould also made a considerable name for himself at home, finally attaining the scientific status he had yearned for as an ornithologist, on or off the field. He was now a professional. In 1843 he was elected a Fellow of the Royal Society, the greatest honour that could be conferred upon a scientist, and one rarely given to a zoologist.

His new-found reputation was also reflected in smart London society flocking to see the latest novelties from Australia – dead and alive – in the ornithologist's house in Broad Street, as well as the number of subscribers willing to pay for his extravagant Australian publications, which would provide him with an income for the next 30 years. Gould had struck the last ornithological goldmine.

CHAPTER 10

The Sacrifice to Science

GOULD RECEIVED a hero's welcome from the scientific fraternity in Europe for his work in Australia, but few spared a thought for the 'most able coadjutor' who had helped to make it all possible, and who was still slaving away in the bush in pursuit of ever more specimens to add to Gould's collection. John Gilbert's contribution to the *Birds of Australia* was largely taken for granted, not least of all by John Gould himself.

On learning that Gould had left three full weeks before the deadline for their rendezvous, Gilbert wrote a disgruntled letter to his employer:

Sydney   May 4, 1840

Dr Sir:

I hasten to inform you of my safe arrival at Sydney yesterday, that you were disappointed at not seeing me I can very well suppose, but your disappointment could not exceed mine, to find on my arrival here that you had sailed just three weeks for England. You say I could not expect you to wait my uncertain arrival, I must say I thought you

would not have left before the time expired which I stated
you might expect me, vis all April.

He was late, he told Gould, only because he had taken
advantage of an opportunity to stop off at King George's
Sound, a region, he thought, where he 'should find something
different' to the flora and fauna he had encountered at the Swan
River. 'I thought I was doing my best', he complained, 'at the
same time studying your wishes.'

Gilbert had been reluctant to return to Sydney with what he
considered an inadequate collection, apart from the birds,
which contained some of the most remarkable specimens to be
included in Gould's work. Gould wrote to Gilbert in
November 1840:

> 'Your collection did contain a few new and very interesting
> birds of several of which I regret to say there were only a
> single specimen; for instance the lovely little Parrakeet
> which you have called Nanodes pulchellus I never saw
> before. It is a little beauty. If you meet with it at Port Ess.
> get all you can quietly and also of the Rock Nanodes. I
> wish you had sent more of this bird. The Brown tree
> Creeper is also new; but of all the birds in your collection
> the one that gave me the greatest pleasure was the Native
> Pheasant which I have published in the first Part, with
> your notes . . .'

Before Gilbert's arrival very little collecting had been undertaken in Western Australia, and that mainly by a few French
voyagers and some of the colony's early explorers. John Hutt,
the Governor, had written to Gould expressing his pleasure
that the eminent ornithologist had seen fit to investigate, albeit
by proxy, the natural history of his own unique, yet sadly
neglected, colony: 'Much of the knowledge of this part W. A. is
based on ignorance. The few passers by who have pretended to

any skill in listing [?] or otherwise, seem to hint that we have mistaken resemblance for reality. The trees and plants for instance are called by the same names as those of Sydney or V. D. Land; same things with the same propriety as a Bird is called a Laughing Jackass.' Gilbert's exploration was the first scientific investigation so far.

In the end Gilbert's collection contained (as he records in a letter to Gould of 25 February 1840) at least 175 species and 750 specimens of birds, 80 species of eggs, 100 skeletons, and 18 species of quadrupeds. In addition, he had, as we know from a letter of the previous October, at least 500 insects, 400 shells, a few crustacea, 7 bottles of reptiles and about 400 plants. And yet he expected Gould to be disappointed. The diversion to King George's Sound was a last attempt to add more quadrupeds to the list. (Gilbert did, at least, add a species of rat kangaroo to his collection while in King George's Sound. It was described by Gould at a meeting of the Zoological Society held on 9 Feb 1841 and given the name *Hypsiprymnus gilbertii*. The species disappeared shortly after the turn of the century.)

His efforts in Western Australia, Gilbert emphasised in defence of his not-insubstantial collection, had been hampered by a severe lack of proper equipment, of the necessary books of reference (the only one he had to hand, naturally, was Gould's *Synopsis of Birds of Australia*), lack of assistants, and a shortage of money. It was not his fault if his collection did not come up to scratch, as he wrote to Gould:

'While I was at the Swan, I was sadly deficient in almost every thing necessary for collecting except Birds . . . sending me as you did without the necessary things and also crippled in a pecuniary way & in a most expensive country, & where nothing that I was in want of could be procured. It is with feelings of vexation to myself individually that I have sent home so poor a collection of Natural History generally, but at the same time I must

inform you that Western Australia is wanting in extraordinary novelties (Quadrupeds at all times and places
may be said to be unabundant) as compared with other
parts of the Continent, Capt. Grey who has just returned
to England saw a good deal of me & I trust will give me
credit for being as industrious in the Colony as anyone
could be you must remember I had none of your
advantages, what could I not have done with your Bullock
Cart, Horses & Men? the whole of my collection has been
formed by myself unaided by a single person during my
whole stay in the Colony . . .'

Gilbert's anxiety about Gould's response to his collection says
something about the demanding standards and the high level of
achievement Gould expected from his collectors – as well as
Gilbert's desperate desire to please. Gilbert wanted to assure
his employer that, if he had fallen short of expectations, it was
not from want of effort on his part; he had laboured no less
intensively than Gould himself, as many of his contacts
testified.

One of the settlers with whom Gilbert stayed, a Mrs Robert
Brockman, who had a house in Guildford in the York District,
described years later how Gilbert used to spend his days while
he was collecting in Western Australia. He was clearly a man of
great charm, and generated great affection:

'We liked him so well, that we told him to consider our
house his head-quarters whenever his occupation brought
him within reach of us; and he was a great deal with us
while after the birds he was in search of.

He used to go out after breakfast, provided with some
luncheon, and we seldom saw him until late in the afternoon, when he would come in with several birds and set
busily to work to skin and fill them out before dark. In the
evenings he used to sing for us, and it was a great treat to

hear his lovely voice, for such a beautiful tenor voice was
rarely heard in those days. He had a good collection of
songs. The opening verse of one of his favourites was:—

'No more shall the children of Judah sing
The lay of a happier time,
Or strike the harp with the golden string,
'Neath the sun of an Eastern clime.'

He was an enthusiast at his business, never spared
himself, and often came in quite tired out from a long day's
tramp after some particular bird, but as pleased as a child
if he succeeded in shooting it . . .

I remember his face now perfectly, as he used to look
when he came in and threw off his heavy pack. He would
say: "Now for a sup of your nice tea and I shall be all
right."'

The Western Australian stage of his journey now over, Gilbert
had a mountain of matters to discuss with Gould: the achievements and failures of his trip, his problems with money and
equipment, and his plans for the future; quite apart from need-
ing the personal reassurance of seeing old friends and using the
opportunity to pass on messages for people in England.

It would have been difficult for Gilbert to take Gould's
departure as anything but a hurtful lack of concern for him. But
then Gould's letter to Gilbert, hastily written in his most
illegible hand, added insult to injury. His instructions were at
best 'rather indefinite'; Dr Bennett, too, complained that
Gould 'did not explain things as particularly as he could have
wished.' Gould had given Gilbert a 'very long list of
Desiderata' – of birds he personally had not been able to find –
with only the difficult scientific names for reference. As Gilbert
said:

'You must be aware that a very great number of these
Birds I have never seen & unfortunately for the want of a

Knowledge [sic] of Latin I must remain in utter ignorance,
of a very great many of them, but each of them as are to be
seen in the Museum [in Sydney] I will endeavour to make
myself acquainted with. . . . I shall not withstanding this
be as zealous in collecting everything I meet with & study
their habits, as if I knew them, but still very little trouble
and a few words in your list would have made me
sufficiently acquainted with each species to recognise if I
see them. I wish too that you had left me a double barrelled
gun, I can assure you I have lost many good Birds for want
of a second shot.'

It seems extraordinary that Gould could have forgotten
Gilbert's lack of education, scientific or otherwise. Their
backgrounds were so similar that one would expect Gould to
feel some sympathy for Gilbert's inadequacies. It is incredible
too that Gould should have left his most valuable collector with
inadequate equipment. Could there have been some reluctance
on Gould's part to allow his assistant too much success? Gould
was still desperate for new species, and he exhorted Gilbert to
do his best to find them, but he was not going out of his way to
make it easy for him. Gilbert saw this behaviour as part of
Gould's self-preoccupation. The oversights of Gould's letter
were put down by Gilbert to 'I suppose, the hurried manner
you must apparently of [sic] written'.

Gilbert also endured Gould's parsimoniousness with fortitude. Ever the meticulous housekeeper, Gould expected
Gilbert to keep a detailed account book, recording all his
personal and professional expenditure, of which the former was
to be repaid by Gilbert on his return to England. The amount
Gould had given Gilbert in Van Diemen's Land had barely
covered his costs while he was in Western Australia, but
Gilbert valiantly sacrificed his own personal comfort for the
sake of professional success. 'While I was at the swan', Gilbert
wrote, 'I Debarred myself of many things, I stood in great need

of that I might throw as much of the small sum you entrusted me with into your service.' Gould was well aware of the extortionate price of commodities in the colony, let alone in the more remote settlements, and should have acknowledged Gilbert's unselfish thriftiness. Instead he beleaguered him with constant financial warnings. 'You still remind me of the necessity of being economical', wrote Gilbert patiently, 'I shall not lose sight of this instruction.'

Gilbert did show signs of irritation, however, when he discovered where Gould had expected him to stay while he was in Sydney, and that all his personal belongings, which he had left in Gould's care, had been stolen. This was the second time Gilbert had been the victim of theft, and it must have been particularly galling that he had lost the last of his few luxuries when, after months of self-deprivation, he longed for them most. In a letter to Gould of 8 June 1840 Gilbert wrote:

'The Tent you spoke of as being the residence of James (and which you seemed to hint I might make as mine) I found in such a wet & filthy state, I would not even have slept in, in the Bush, besides which I could never have left the place, more particularly after examining my trunk (which I understand was kept there). When you hurried me from Launceston [in Tasmania], without my getting my things, you promised to keep the key in your own possession, but I am sorry to say I learn from Mr. Bennet [sic] you did not do so, the consequence is, I have lost a very considerable part of the contents I left in the trunk. In fact, the whole of my things have been fairly turned over and all the most valuable portions taken out, a list too that I had written, after being robbed at Hobart Town, is gone so that I cannot tell the full extent of my loss, all my Blankets – my Pillow – Sheets – Towels – Pillow cases have been taken away, with several little trinkets, the intrinsic value of which was certainly not great, but to myself the loss is irreparable. I hope I shall not offend you in writing so much of my loss, and I beg to be understood as attaching to no person in particular the blame of having made free with my property, but you must suppose – when I had no alternative but [to] leave my things in your charge, that it is extremely hard and annoying to find no care or thought has been bestowed during my absence.'

Gilbert was completely devoted to Gould's cause, and it can only be a credit to his employer that despite the oversights and insensitivity he displayed on a personal level, Gould managed to inspire in Gilbert an enthusiasm and dedication almost equal to his own.

Gilbert was a reliable collector in other branches of natural history as well and Gould was keen to take advantage of his assistant's abilities in every field. Among Gould's lengthy and detailed instructions were sweeping requests to 'collect all the Insects you can', and, 'Pay as much attention as you can to Shells, particularly land shells'. There was no area of zoology Gould did not expect Gilbert to cover.

Gilbert managed to secure a passage to Port Essington on the north coast – the main hunting-ground still at the top of Gould's agenda – on 15 June, sooner than he had expected. A severe hurricane had hit the isolated outpost, and a government ship was dispatched from Sydney to relieve the stricken settlement. 'You would have been highly amused', Gilbert told Gould, 'if you could have witnessed the impatient anxiety of the people to get the first drinks of Beer or Spirit that could be spared from the Ship.'

Gould's suspicions that Port Essington might prove another eldorado were confirmed by Gilbert the moment he landed. On 18 June he wrote, 'I have been out once, and saw so many beautiful species of Birds that I knew not which to shoot at first . . . I shall be enabled I think to form as splendid a collection as we have done in any part of Australia.' While at first sight the vegetation seemed not unusual, the birds, Gilbert said, were generally different to any he had ever seen before. He could hardly contain his excitement: 'I imagine the greatest Australian novelties now are to be obtained here . . . I shall be very much disappointed if I do not produce in my collections something new for your great work.'

Three months later, on 19 September, Gilbert wrote triumphantly, 'I have collected ninety Species of Birds, six of which are included in your list of desiderata.' Over 30 of the remaining were new, at least to Gilbert. All together, he wrote, 'I have already above two hundred specimens which I think will satisfy you.' He had made one ten-day trip into the interior on which he 'went several miles further than any European had ever before attempted' and spent a week exploring the islands in Van Diemen's Gulf. 'I trust', he ended, 'I shall astonish you with my collection, as I am devoting the whole of my time to it to render it valuable in species and beautiful in preparation, of course I made Birds my first study, and although everything else is secondary, I am happy to say I can find sufficient time to devote to other subjects.'

Gilbert's jubilation was met with a typically cool response from London. There were no congratulations, little encouragement, and a stern reminder to make haste with his dispatches. 'I received both your Letters from the latter place [Port Essington]', wrote Gould on 5 April 1841, 'the first announcing your safe arrival and the other bearing the date of 19th September 1840. I was of course ['glad' (crossed out)] happy to hear that you were in good health and that you were going on ['favourably' (crossed out)] so well: I now wait with some degree of impatience the arrival of at least a portion of your collections as knowledge of the productions of the north coast is most necessary even in this early stage of my work.' Because, Gould wrote, of the 'enormous expense of my expedition' Gilbert was once again required 'to make the expense as little as possible', and to consider Port Essington his last stop. Gilbert needed no reminder. He had already sailed for Singapore on 17 March taking his valuable collections with him. By the end of September he had reached London.

He found Gould with four parts of the *Birds of Australia* and the first part of his *Monograph of Kangaroos* already published. By now Gould had a precise idea of the specimens he was lacking, and, gratefully receiving Gilbert's stunning contribution, he urged his collector to return immediately to the field. Gilbert was only four months in London before he set sail again for Australia, this time 'for the term of at least 3 years'. He was supplied with a formal letter of explanation from his employer for use on his trip:

> London, 20 Broad Street
> Golden Square
> Jan. 21, 1842

Dear Sir

The fact of my directing your return to Australia for the purpose of still further investigating the zoological productions of that vast country is the best proof that I have been satisfied with your labors both during the time you were with me in Van Diemen's Land and since we separated in that country, you for the western and I for the eastern shores of Australia.

Few expeditions, whether undertaken by private individuals or governments, have terminated so happily as our own, and certainly none have been productive of so large a number of novelties in the field of science or of so many singular facts relative to the economy of the subjects discovered as our joint expeditions have been, and I sincerely trust that you return to that fine country with the same ardor and with the same persevering and industrious habits that signalised your conduct upon the former occasion.

Believing such to be the case I have only to wish you the

blessing of health and prosperity in all your engagements.
I am,
Dear Sir
Yours very truly
John Gould.

The letter was accompanied by a contract, drawn up by Gould, signed by Gilbert, and witnessed by Prince, which stated that 'The objects of this your second journey are twofold; the first and principal of which is to collect the zoological productions of the countries you are directed to visit together with all information respecting their habits and economy, the second is to endeavour to obtain subscribers to my publications and to collect subscriptions for them.' All Gilbert's collections and observations, the contract ordained emphatically, were to be transmitted 'for my sole use'. Should Gilbert take advantage of a passage on one of Her Majesty's vessels, he should first 'come to an understanding' with the Captain that 'The specimens you may collect are expressly for me'.

A separate long letter of instructions vested Gilbert with some challenging tasks:

'Collect skins of *all* kinds of kangaroos in New South Wales and other mammalia, with their crania; send, if possible, the nose, face, and the palms of the hands and feet of all kinds of the smaller animals in brine or spirits'; 'Gain if possible information respecting the nidification of the Cuculidae. Ascertain if possible positively what species lays the olive-brown eggs, by searching for the eggs in the body'; 'Procure the wallaby of Garden and Rotnest Islands. Dissect as many female as possible and ascertain if the young are found in the uterus and pouch at the same time; state the size of the young in the pouch'.

In addition to his beloved birds, Gilbert was expected to send

Gould fish from Western Australia and New Holland; as many shells as possible; plants, ornamental shrubs, and pieces of bark for drawings; all the sponges and corallines possible; reptiles and insects; the skins and crania of all quadrupeds, great and small; even some live specimens for Lord Derby's menagerie; and, obscurely, the beard of the *Pinna* from the Sound.

While New South Wales was not to be a main focus of Gilbert's attentions, he was asked to visit Natty and Jemmy at Yarrundi to try and solve the nagging questions that had been preying on Gould's mind ever since his visit there – the nidification, or nest-building, of the lyre-bird being one of them. There were loose ends, too, to be tied up in Western Australia and a second visit to the remarkable region of Port Essington, as well as a journey to the fresh hunting grounds of Moreton Bay, in the region of Brisbane.

On the 1 February 1842, Gould bade his assistant a curt and final 'bon voyage' aboard the *Houghton le Skerne* at London docks. It was the last time Gould would ever see him. Four days later Gilbert wrote a dejected if humorous letter to his employer from the ship, explaining that neither his departure nor the company on board were as propitious as both of them had hoped. Gould's last words to Gilbert had been, 'If the vessel dont sail I would not come back again if I was you.' The vessel did not sail, at least for another 24 hours, and Gilbert, effectively banned from reappearing at Broad Street, was forced to spend the day amusing himself as best he could. It was, however, the company on board ship that Gilbert found most objectionable:

'I have heard it remarked that a feller may be passenger during his life in fifty vessels, yet he shall not meet two crew similar this remark has been fully borne out in the present instance with myself for this is the strangest vessel and the oddest people to manage her I have yet seen, you know from the specimen you saw of the Captain and his wife that they are homely people but I am sure you could not any more than myself imagine the extent of it. The first [illeg] was a sickening specimen, a Table cloth that was all colours, and would puzzle any one to tell whether it had been originally white blue or brown, and the soup brought on the table in a common yellow pie dish, I thought this was pretty well, but this was not all nor was I (even after this) at all prepared for what followed. I went on deck after dinner to smoke a cigar, and on returning below a little time after in the cuddy, judge of my surprise and horror, not immingled with even a greater share of disgust, on beholding Mrs. Proud sitting by the fire *smoking a pipe of strong tobacco*, if there is one habit that we poor human beings give way to that is worse than another, I think it is a *Woman smoking*, at all events you will say with myself it smacks very much of Billingsgate the poor woman I believe was rather unexpectedly caught luxuriating in her favourite cloud, and my surprise no doubt was so plainly expressed, that she thought some apology necessary, and her excuse was, that she was subject to fits, and smoking the strongest Tobacco she could obtain prevented their attacking her so often, the great objection and annoyance of this habit is that when she sits at the Table we fancy she smells strongly of Tobacco, when poor woman *perhaps* she may be quite innocent, but by far the worst part of the matter is, that she acts as stewardess, and makes all the pastry, I fear I shall have very little taste this passage for Puddings & Pies. At the sight of either I fancy they were made while she had a pipe in her mouth . . .'

At last, however, a fair wind blew for the *Houghton le Skerne* and the ship was finally underway for Freemantle, Western Australia. Once on course, despite a general attack of sea-sickness, Gilbert's optimism returned. 'I am rejoiced to say', he wrote to Gould on 7 February, 'I feel in excellent spirits, dreaming every night of Kangaroo's [sic] and Native Pheasants.' With the consoling prospect of Australia ever in view, Gilbert survived the rigours of life on board the ship with Mrs Proud and her pipe, and arrived in Fremantle, after a stay of five weeks in the Cape, on 17 July 1842.

He immediately resumed his work, teaming up with James Drummond, the Government botanist and an old acquaintance, 'he to collect Botanical Specimens, & I to scrape together a sort of omnium gatherum'. Drummond, who had emigrated to the Swan River colony in 1829, had a station on the Moore River, and it was from here that the two made their first expedition. The Drummonds became Gilbert's close working partners, and James' son, Johnson Drummond, who took a particular interest in Gilbert's work, was trained with a view to becoming Gould's principal collector in the area after Gilbert had left. Gilbert told Gould:

'From Johnson Drummond I have obtained a vast deal of information, which will render the history of many Birds & Animals most interesting. This young man knows almost every production of the Colony, most fortunately I have tried him in many instance, & invariably found his descriptions (most minute in particulars) agreeing perfectly with my own observations, he has been telling me lately of two Birds, & which he has promised to procure for me, which, from his descriptions must be quite new . . .'

A trip to the Wongan Hills with the Drummonds in September provided Gilbert with the invaluable opportunity of observing the habits of the mound-building mallee-fowl, and his description of this extraordinary bird, sent to Gould on 28 September, was read at a scientific meeting of the Zoological Society on 13 June the following year.

Gilbert's third excursion with James and Johnson Drummond to the area around Augusta yielded Gilbert's rarest and most significant ornithological find – the so-called noisy scrub bird, a species that Gilbert identified, according to Drummond, as being 'allied to the Australian bristle-bird, but without bristles'. (This bird, believed to have become extinct before the turn of the 20th century, was rediscovered in 1961, since when it has been bred on a special reserve and closely scrutinised by scientists. It is thought to be one of Australia's most primitive birds, dating back 40-55 million years, and therefore reveals vital evolutionary evidence about the early radiation of Australian song-birds.) 'It was only after many days of patient and motionless watching among the scrubs', says Gould, 'that he succeeded in obtaining specimens.' The extraordinary bird was exhibited by Gould at a meeting of the Zoological Society in London on 9 January 1844 and later in his *Handbook* Gould described it precisely according to Gilbert's description: 'The total absence of vibrissae (bristles) in a bird apparently closely allied to *Sphenura* renders it one of the anomalies of the Australian fauna.' The observation, though, as remarkable as it was, was not attributed to Gilbert.

Meanwhile, Gould was not proving reliable or encouraging in his back-up from the Broad Street base. Gilbert was indignant. He was, once again and all too soon, forgotten. On learning that a ship from London had arrived in Perth, Gilbert rushed to meet it:

'You know with what delight one hastens from the Bush in expectation of reading Letters from far distant Home, & you may readily guess how much I was mortified on beholding half a dozen lines only from you . . . I do think it is very hard, & particularly unkind in *some* People not writing by the "Simon Taylor", which did not sail till nearly four Months after my Departure from England, & a vessel leaving London with *200 Emigrants* must have been frequently noticed in the Papers; so that ignorance of her sailing can form no Excuse; all I can say is I felt

> extremely vexed at riding as hard as I did Sixty Miles &
> only to receive a short Note . . .

Gilbert's remaining 17 months in Western Australia were spent on a 'perilous trip' to the Abrolhos – a group of remote islands in the Indian Ocean, 50 miles from the mainland and 200 miles north of Perth – from which, tossed about in a tiny boat in treacherous seas, he was lucky enough to secure a 'miraculous escape'; and a second, overland journey to King George's Sound about which little is known. All the specimens Gilbert obtained during the first six months of his work he dispatched to London in HMS *Beagle*, which left Freemantle in May 1843. The remainder were sent that December in the *Napoleon* before Gilbert left the colony for good. Altogether the two shipments contained 318 mammals and 432 birds, plus the eggs of 77 species of birds. Including his first visit, Gilbert had spent a total of two years and four months in Western Australia, nearly a year longer than Gould had spent on his entire trip to the continent.

February to August 1844 found Gilbert on the opposite side of the continent scouring New South Wales for Gould's desiderata and travelling northwards as far as the Darling Downs. Gilbert found several new mammals and a 'totally new parrot' – 'without exception', he considered, 'the most beautiful of the whole tribe I have ever yet seen in Australia'. In an uncharacteristic appeal, he wrote to Gould on 8 June 1844 asking, 'If you have not already honoured my poor name in your works, I know of no species that wd. delight me more to see Gilbertii attached to than this beautiful Bird.' This was the first description of the paradise parrot and was named by Gould, despite Gilbert's plea, *Psephotus pulcherrimus*. It is now believed to be extinct.

In general the area around Moreton Bay proved too similar to the region already explored by Gould further south to provide any dramatic new finds. Gilbert was hungry for greater glories, something, as he put it, 'as will give an eclat to the conclusion of Mr Gould's splendid work.' Suddenly, opportunity knocked at his door.

In a letter of 14 July 1842 Gould had warned Gilbert about the complications of venturing on a government-funded expedition. 'Undertake no journey into the interior', he had said, 'with Government expeditions unless you previously have it stated in writing that all the specimens and notes you may collect are for my sole use and to be made known to the scientific world through the medium of my publications.' A 'small private expedition' led by the German, Ludwig Leichhardt arrived in Moreton Bay in August 1844 to embark on an overland journey to Port Essington. The company was a motley assemblage of Leichhardt, a short-sighted, unreliable explorer *manqué* with only one rather aimless adventure to his name; a 24-year-old Englishman by the name of John Roper, a 44-year-old convict called William Philips; a ship's boy of 15 called John Murphy; the 19-year-old James Calvert; and two Aborigines – Henry Brown and Charlie, a native policeman in Brisbane. They were perhaps not the most likely party to attempt the first overland journey around the Gulf of Carpentaria.

Although there had been speculation for years about the potential of the unknown country lying between the two settlements, it was a venture never before attempted. The expedition, covering a distance of over 2,000 miles, was expected to take five to six months. It took well over a year. For Gilbert the project promised 'a glorious opportunity of unmasking the hidden novelties of tropical Australia', and the expense – under £60 for the whole trip – was not much more than the cost of a passage by ship to Port Essington. Even Gould, Gilbert thought, would be satisfied with that.

Gilbert described the irresistible attraction of joining this expedition to Dr Bennett in Sydney: '. . . when I have penetrated but comparative short distances beyond settled districts', he said, 'I have been rewarded with many novelties, and I must confess it is this which spurs me on at this moment, at the same time when I first heard of the German's intentions, I felt a little jealous of a foreigner being the first to make known the hidden treasures of this vast and interesting country, which has been for so many years the particular province of our own countrymen.' It was admittedly a 'very hazardous under-taking', but, he added philosophically, 'I have fully made up my mind to meet dangers or hardships, both of which we must all expect will more or less be our share before we have accomplished our objects.'

Gilbert's explorations of the region around Port Essington had relieved him of many of the anxieties he had suffered in the past. Unlike Gould, he distrusted and feared the Aborigines, and this had hampered much of his work in Western Australia. He had once written to Gould from the Swan River, 'I have been in the Interior as far as any Europeans have yet settled, but at the time I was there, the Natives committed several frightful murders on the white people, who to punish them killed several of the Blacks in return, this had roused up all their former savage disposition and it was considered very dangerous to move far away from the settlers houses . . .' His anxiety was no doubt fuelled by making the acquaintance of Lieutenant George Grey – later to become Governor of South Australia and one of Gould's many corresponding collectors – who had been attacked and wounded by Aborigines on his expedition from Shark Bay.

At Port Essington Gilbert was relieved to find the situation very different. 'I am happy to say', he told Gould, 'the Natives are extremely quiet, and on very friendly terms with the English, so much so that I felt less fear in going about the country already with the imperfect knowledge of them than I did at the Swan after becoming perfectly acquainted with their habits and manners, to the time of leaving that part.' Secure in the knowledge that the natives were docile, around Port

Essington at least, Gilbert was optimistic about the expedition's outcome. 'I freely confess', he told Bennett, 'I feel as sanguine as others of the party of ultimately reaching our destination.'

Gould was thrilled with Gilbert's initiative, and wrote immediately he received the news of the imminent expedition to Dr Leichhardt, 'Whatever may be the result of the enterprize you have so boldly undertaken that of crossing the continent of Australia you have my constant prayers for the success and safety of yourself and party as well as my warmest thanks for permitting Mr. Gilbert to join you.' Gould was also determined, however, to ensure that no new specimen should slip through his net:

'I should be glad if you would see it. [sic] If in the course of the expedition you have so spiritedly undertaken you should have procured any species that Mr Gilbert does not possess, or if in any future expedition you should do so, May I ask the favour of your sending them to me for the purpose of describing and figuring together with any notes respecting their Economy the specimens should be very carefully preserved for you, or after figuring them I would transmit them to any person in Europe to Whom you might direct me to forward them. You have too noble a mind I am sure to take this in any other light than a right one or to attribute to me any other motive than that of an ardent desire to add to our knowledge of that department of science to which we are mutually attached . . . Your journey is exciting considerable interest in England, and everyone is anxiously waiting to hear of your having arrived safely at Port Essington.'

The motley band disappeared over Darling Downs on 18 September 1844 and the world waited with baited breath for it to reappear on the other side of the continent. The time of their scheduled arrival came and went with no sign of them. By July 1845 Gould was reassessing the expedition he had applauded the year before. It was, he said, a 'most perilous undertaking'. By September 1845 still no news had been heard, and Prince wrote to Dr Bennett expressing his concern for Gilbert's safety. The occupants of 20 Broad Street waited, he said, 'in daily expectation that our worst fears may be realised'. Something terrible must surely have happened. Rumours began to circulate, involving stories of an Aborigine attack. On 16 December 1845 Gould wrote to Fairfax, his agent in Australia, requesting him to trans-ship 'the cases left in your charge only in the event of the sad report that has reached me of [Gilbert's] being murdered being confirmed'.

In the midst of all this anxiety arrived the news of the death of Gilbert's friend and colleague Johnson Drummond in Western Australia. 'I have just learnt', Gould wrote to Bennett in Sydney, 'that the person whom he [Gilbert] engaged to collect for me in W. Aust. has been barboursly [sic] murdered by the natives.' There was little doubt now in Gould's mind that Leichhardt and his part had met with the same fate. Accordingly Gould asked Bennett to collect together 'any letter or papers you may have' belonging to Gilbert 'that the same may be given to his family if necessary'. Gould also wrote to Mr J Stokes requesting that the balance of payment still owing to Johnson Drummond be made out to his father. Of the Leichhardt expedition, he wrote, 'none of them have since been heard of still I hope for the best'. On 1 April 1846 he passed on the same sentiment to Sir William Jardine: 'Nothing yet heard of Gilbert or party but I am still in hopes of them coming out of the country again someday or other.' For the past year Gould had relied on the enthusiastic if somewhat unreliable contributions of a young uneducated collector named Frederick Strange. Gould was reluctant to hand over the reins of such an important job to this unpredictable young hothead, and hoped for his 'most able coadjutor' to return.

Suddenly, the long-awaited news arrived. 'A report has just reached London', wrote Gould to Bennett on 1 June 1846, 'that Leichardt [sic] Gilbert and party have arrived at Port Essington. I fear it is too good to be true but I do not yet despair of their safety and this it is that has prevented me from acceding to Strange's solicitation to appoint him Gilbert's successor.' The story though, as it was relayed to Lord Stanley by Governor Gibbs in a long letter of 29 March 1846, was not the one Gould was waiting to hear. The Leichhardt party had indeed reached Port Essington on 17 December 1845, in a state of extreme exhaustion and suffering from exposure and starvation; but Gilbert was not among them. Six months earlier, on the Cape York Peninsula, the party had been attacked by Aborigines and Gilbert had been killed. Gilbert's worst fears had been realised in that part of the continent where he had least expected it.

The full circumstances of the murder were not made known to Gould until that autumn, when he received a letter from a member of the party, John Roper, giving an eyewitness account of the attack. It was read out to a hushed meeting of the Zoological Society in September:

Sydney, May 12, 1846

Dear Sir, – As I was one of the party that journeyed from Sydney to Port Essington, and not knowing whether you had been made acquainted with the full particulars of poor Gilbert's death by Dr Leichhardt, or any other of the party, thinking the details of his melancholy fate would be read with interest, I shall offer no apology for addressing this to you.

As Mr Gilbert's log, which has been sent home to you, fully narrates all particulars up to the eventful 28th June, I shall offer no remarks of my own. At the most northerly point we reached on the east side of the Gulf of Carpentaria, in lat. 15 57′, and about fifty miles from the coast, we encamped for the night at a small shallow lagoon surrounded by low tea-trees, the country around beautifully open. Having partaken of our usual meal of dried meat about 3PM. Gilbert, taking his gun, sallied forth in search of something new – he procured a Climacteris and a Finch, which he skinned before dinner: our scanty meal was soon despatched; poor Gilbert was busily employed plaiting the cabbage-tree, intending to make a new hat, which, alas! he never lived to finish. The shades of evening closed around, and after chatting for a short time we retired to our separate tents – Gilbert and Murphy to theirs, Mr Calvert and myself to ours, and Phillips to his; the Doctor and our two black fellows slept round the fire, entirely unconscious of the evil designs of the natives having always found those we had passed so friendly and well-disposed, we felt in as great security as you do in the midst of London, lying on our blankets, conversing on different topics. Not one, I think, could have closed his eyelids, when I was surprised by a noise, as if some persons were throwing sticks at our tent; thinking it must be some trick played on us by our companions, I sat up to look out; another volley of spears was thrown; a terrific yell, that will ring in my ears for ever, was raised, and pierced with spears, which I found it impossible to extricate, I sunk helpless to the ground; the whole body rushed upon us with their waddies, and how it is it that our brains did not despatter the ground is to me miraculous. These rascals had crept on us under cover of the tea-trees: the tent in which Calvert and I were being first in their road, the whole body attacked us; poor Gilbert, hearing the noise, was rushing from his tent with his gun, when a spear thrown at him pierced his breast, and, penetrating to his lungs, caused haemorrhage; the only words he spoke were these, 'Charlie, take my gun; they have killed me', when pulling the spear out with his own hands, he immediately dropped upon the ground lifeless. Little Murphy, who was by his side at the time he was speared, fired at the black fellow who speared him; Brown fired at the mob beating Calvert and myself, and they immediately retreated howling and lamenting. Mr. Calvert was pierced with five spears, myself with six, and our recovery is to be attributed to the abstemious way in which we lived. After having the spears pulled out, you may imagine our feelings when we heard Charlie exclaim, "Gilbert is dead!" – we could not, would not, believe it. Alas! the morning brought no better tidings – poor Gilbert was consigned to his last and narrow home; the prayers of the church of England were read over him, and a large fire made upon his grave for the purpose of misleading the blacks, who, we thought, would probably return and search the camp on our departure. It is impossible to describe the gloom and sorrow this fatal accident cast upon our party. As a companion none was more cheerful or more agreeable; as a man, none more indefatigable or more persevering; but it is useless for me to eulogize one so well known to you – one whom you will have cause to regret, and who will ever be remembered by,

    Sir,

Yours most truly,

John Roper

Despite, or even because of, the death of Gilbert, Leichhardt's expedition was given a triumphant reception in Sydney. Leichhardt was hailed as a hero who had brought his company through every trial and tribulation with the loss of only one life. His account of the expedition did much to glamorise its author and publicise his achievements. Only those with an intimate knowledge of the bush and its inhabitants must have been sceptical about the truth of Leichhardt's account. Gould, who had a greater understanding of the Aborigines than most, might have had his doubts. His friend Captain Sturt, who arrived back in Adelaide from his valiant but fruitless expedition into central Australia only a few weeks before Leichhardt reached Sydney, might also have nursed his suspicions. Sturt had not experienced one moment's hostility on his journey, despite crossing the territory of reputedly the most dangerous tribes in all of Australia. His expedition, however, received hardly a fraction of the public acclaim that Leichhardt's did.

No official questions were asked of Leichhardt, and he seemed confident that none of his company would cast aspersions. Shortly after his return to Sydney, however, he sent Gilbert's pocketbooks to Gould, believing presumably that they contained only ornithological data. In fact they contained relatively little information on birds, but did contain a damning refutation of Leichhardt's character and conduct as leader of the expedition. Gilbert's testimony, written daily and with considerable accuracy, implied that Leichhardt, far from being the archetypal leader he made himself out to be, was hot-tempered, irresponsible, irrational, arrogant and dishonourable.

Gilbert's journal also gave a very different reason for the Aborigine attack than that presented by Leichhardt. The day before he died, 27 June 1845, Gilbert described how he suspected the two natives from the Leichhardt party, Charlie and Brown, of committing some atrocity against a local tribe nearby, perhaps even of raping one of their women: 'As I was returning to camp reports of the guns of Charlie and Brown were heard in the direction I had just come from, and immediately after we heard the shouts of numerous Natives; and when Brown and Charlie returned soon after they told us they had caught the Natives in the act of creeping up to our bullocks with their spears ready poised to kill one. They ran after them and fired off their guns to frighten them off, in which they succeeded.' Their story, however, did not wash with Gilbert, whose entry continues: 'I am inclined to think that the real cause was that our blackfellows surprised them at their camp, and as I know Charlie would not be very particular in his treatment of a native woman if he caught one, it seems to me that the men perhaps resisted, when the boldness of our fellow in the superiority of their weapons effectually drove them off . . .' Hours later Leichhardt's camp was attacked and Gilbert was dead.

No mention of Gilbert's suspicions appears in Leichhardt's published account. The leader of the expedition simply accepted the story as told to him by Charlie. Even when two months later Charlie and Brown quarrelled and one of them confessed that they had assaulted three women and wounded an old man on the fateful night of 27 June, Leichhardt refused to acknowledge Gilbert's death as anything but a hostile and unjustified attack. To the other members of the expedition it was all too clearly a retaliation; as Leichhardt continued to encourage the wanton and rebellious behaviour of Charlie and Brown, they feared that their own lives might fall prey to a similar act of retribution. That they made it to Port Essington without further incident under the leadership of a madman is a miracle. But the voyage had taken nearly ten months longer than expected, and some suspicions must have been raised, among other explorers and scientists at least, about the effectiveness of Leichhardt's leadership.

The truth lay smouldering with incrimination on Gould's desk in Broad Street. But we hear nothing from the ornithologist on the subject of Gilbert's death other than a few words of regret and a stiff acknowledgement in *The Birds of Australia*, designed to absolve Gould of all responsibility. 'Unfortunately for himself', Gould writes in his preface, Gilbert 'allowed his love of science, in the advancement of which no one was more ardent, to induce him to join Dr. Leichhardt in his overland journey from Moreton Bay to Port Essington.' That Gould had pressured Gilbert to find ever more exotic and plentiful supplies of new species, to go further than any other European had gone; that he had stipulated that one of Gilbert's top priorities should be to explore the very country where he was killed – 'With respect to the Gulf of Carpentaria you know the importance', Gould had written in a letter of July 1842, 'and you will not fail to avail yourself of the opportunity of visiting those parts' – and had spoken admiringly of others who risked their lives in the service of science, was suddenly of no significance. While Gilbert was alive he was Gould's collector; as soon as he died he was, and always had been, his own boss.

Gould's lack of curiosity about the circumstances of Gilbert's death – given the strange and moving testimony that had been surrendered to him, and despite receiving several letters from an Australian collector and friend of Gilbert's that incriminated Leichhardt – verges on the callous and hard-hearted. If anyone could have had the authority to instigate an investigation and to bring Leichhardt to justice, it was Gould. It is possible that Gould did not read those parts of Gilbert's journal that did not contain information about birds; but it would be a remarkable feat of indifference for anyone not to read the last entry in the journal of a close colleague written the day before he died. Moreover, it seems that Gould scoured Gilbert's closely written script with a fine-tooth comb looking for ornithological details, and it is unlikely that he passed over any part of the account. As Gould himself explains, the expedition was far from being a complete disaster now that Gilbert's legacy was in his possession. 'Fortunately', he says in the preface to *Birds of Australia*, 'despite the many difficulties and dangers which beset the party during the remainder of their journey, his [Gilbert's] journals and notes, together with the specimens he had been able to procure, were preserved and transmitted to me by Dr Leichardt [sic], and proved of valuable assistance in determining the range of many of the species.'

The answer to Gould's indifference lies partly in his con-tinuing interest in Leichhardt. He had written to the German in 1844 requesting specimens and observations from the expedition with Gilbert, as well as from 'any future expedition'

Leichhardt might undertake. By 1847 it was apparent that Leichhardt was preparing another even more ambitious project, this time aiming to cross the entire continent from Moreton Bay to Perth. Gould would have wanted nothing – not even the death of Gilbert – to jeopardise potential returns from an expedition of such magnitude, as Leichhardt was well aware. When Leichhardt wrote to Dr Bennett in Sydney asking if he could keep Gilbert's gun, now officially the property of Gould, for use on the expedition, he boldly referred to the ornithologist's vested interests: 'How could Gould ever have an objection to contribute such a trifle to the Expedition' he asked, 'which is as interesting to himself as it is to me?'

Gould's reaction to Gilbert's death is also typical of a man obsessed by his subject to the exclusion of human contact. Gould engaged with Gilbert, as with so many of his collectors and colleagues, on a professional but never a personal level; he had neither the time nor the inclination to make friends of those he considered principally the servants of science. His work was his *raison d'être*; if one of his agents was 'removed', it was not so much the personal loss as the injury to his work that he regretted. Gilbert became another item of expenditure to be considered, along with Johnson Drummond and Sir John Franklin's assistant who had shot himself in the boat in Van Diemen's Land, contributing to the cost, and therefore the value, of his collection. 'It comprised' Gould assessed in the preface to *Birds of Australia*, 'the finest specimens I had been able to procure during the long period of ten years, collected together at an expense of more than £2,000 and at the cost of three valuable lives . . .'

Gilbert's contribution to Gould's opus magnum is barely acknowledged. Gilbert spent five-and-a-half years devoted to collecting in Australia, as opposed to Gould's eighteen months; and visited every one of the present states, while Gould himself saw only the territories of New South Wales and South Australia. He presented between 60 to 70 new species for

Gould to name, of which only one – called unflatteringly 'Gilbert's thickhead' – was dedicated to its discoverer.

Perhaps Gilbert's name was cropping up a little too frequently in Gould's letterpress for the egotistic author. Many of Gilbert's observations from Western Australia and Port Essington, especially for the more unusual species, were unique – his notes still form a vital testimony of many birds that are now extinct or extremely rare – and Gould seems to tire of quoting his assistant in his text. Perhaps also Gould saw no material or political advantage in attributing honours to someone who was safely under contract to begin with and then, later, dead.

CHAPTER 11

A Double Defection

WORK HAD RESUMED at a cracking pace in Broad Street on Gould's return in August 1840. One part of the *Birds of Australia*, each containing roughly 17 plates, was scheduled for the first of every third month from 1 December onwards.

While Mrs Gould, who was pregnant again, resigned herself to the labours ahead, her husband sifted through specimens in the museums in Leiden, Paris, and Rotterdam for further species of kangaroo to incorporate in the *Monograph*. By May there were enough plates completed for the couple to take a break without disrupting the publication schedule, and Gould installed his wife in a 'little retreat' in the country in preparation for her confinement.

The imminent arrival of another child again caused Gould to regret an interruption in his work. 'I fear', he said to Jardine, 'I also shall not be able to go to Plymouth in consequence of Mrs. G. accoutrement [sic] being about that time. I rather regret this as I think I may excite some interest in the Zoological Section by exhibiting full sized oil paintings of the extraordinary bowers made by the Sattin [sic] and Frilneck (Chlamyderma).'

197

It was difficult for Gould to forsake even a minor coup of this sort for matrimonial duties, however he resisted for a time the call of his birds and contented himself with the prospect of visiting Jardine in Scotland after the baby was born, when he would doubtless make up for lost time shooting and discussing his favourite subject. Eliza too was looking forward to a holiday. 'Mrs. G. I assure you', wrote Gould to his friend and patron on 2 July 'looks forward with great pleasure to her visit to Jardine Hall and I trust there will be nothing occurring on her part to disappoint.' Gould's casual and optimistic remark could not have been more cruelly betrayed. Six weeks later, just five days after giving birth to their eighth child, Eliza died. She was 37.

It was left to Prince to break the news to the expectant Jardines, which he did two days later on 17 August in a letter adorned with flowery Victorian sentiment. 'Poor Mr Gould', Prince explained, 'is so overwhelmed with the heavy heavy calamity with which it has pleased the All-wise disposer of events to visit him that he is utterly incapable of writing himself and has therefore deputed to me the painful task of informing you and Lady Jardine of the demise of his dear and amiable wife who was removed from this transitory world of care on Sunday morning at Egham a little before 9am.'

Many accounts suggest that Eliza died of exhaustion. The production of over 500 plates, innumerable drawings (over 600 for *Birds of Australia* alone), and eight children would sap the energy of the strongest woman. But Eliza's death was caused not by lack of strength or stamina – she had had ample opportunity during her 12 years' marriage to Gould to prove herself on both those counts – but by bad medical practice. She died, according to the death certificate, of 'puerperel [sic] fever', or 'childbed' fever as it was more commonly known – a condition often brought on by the doctor, who, not knowing the simple value of washing his hands, could come straight to a delivery from dressing a corpse. 'You will however be happy to know', Gould later told Edward Lear, blissfully unaware of the implication, 'that her own medical man was with her for several days before her confinement and who remained until after her death.' The infection Eliza had picked up killed her within a couple of days.

Gould faced the loss of his wife with remarkable fortitude. 'You will far more readily imagine than I can attempt to describe', said Prince to Jardine, 'the state in which this dreadful visitation has plunged Mr. Gould. I am happy to say, however, that he is much more calm than could have been hoped or expected.' He had his children to think of, and it is mostly on their behalf that he rallied, as he told Edward Lear barely ten days after the event: 'For the sake of my Dear Little Children I endeavour to bear up against the trial as well as I can and I trust that God will give me strength to be the best of fathers to them, an awful charge.' Gould, however, was clearly concerned about how he was to find the time to devote to the services of single parenthood. 'You will my dear Sir William', he said, 'readily imagine how much suffering this very heavy loss has occasioned me and what a heavy duty necessarily devolves [sic] on me in bringing up my six dear little children.'

The solution lay in deciding that his most important duty to his children was to continue his work. The role of surrogate mother he left to the capable hands of Mrs Mitchell, already well versed in the part since the Gould's visit to Australia, and Eliza's kindly and accommodating mother. Later Lizzie recalled of her father: 'We saw very little of him all day, not even at meal times as we had early breakfast & tea in the school room . . . & dined in a front room downstairs next to the kitchen to save the servants trouble, whilst the dining room proper, was considered especially his room . . .' There was little entertaining either, of which the children could partake: '. . . father having no wife did not have ladies to see him, except it was to see the birds & he generally dined at his club while we were little. We always saw him in the morning between nine & ten when he went down to breakfast, & all made a rush at him for a kiss nearly pulling his head off, he used to say, & again when he came up to change his coat for late dinner. Other times he was in his office at the back of the house busy with his books & birds . . .' Occasionally, as the eldest boys 'grew companiable', Gould would take them out on a Sunday afternoon to the Zoological Gardens; but, apart from the odd spontaneous holiday Gould was content to surrender their charge to the other members of the household.

Eliza's death left Gould bereft of a professional as well as a personal partner. Many of Gould's associates, including Prince, could not see how he could continue without her. Charles Sturt told Gould he not only sympathised with Gould's personal bereavement, but 'regretted Mrs Gould as a public loss and brought to mind how difficult it would be for you to replace her.' Prideaux John Selby had expressed the same sentiment to Jardine: 'She will be a grievous loss to him in every way', he said, 'and I should think would almost put a stop to some of his works.'

But they all underestimated the drive and perseverance of the ornithologist. A tragedy such as this served to fuel Gould's determination and inspired him to concentrate single-mindedly on the one subject that could bring him solace. He threw himself with renewed vigour into his work and found boundless comfort there. 'I am happy to say', he told Hugh Strickland in October, 'that I bear up against my loss with more fortitude than I would have expected and I admit that I do feel considerable relief in a close application to my scientific pursuits without which indeed I should be truly miserable.'

Gould refused the advice of friends to leave London for a spell, and even declined another invitation to Jardine Hall. He had at his fingertips a far more efficacious source of consolation and distraction; he was in the middle of the most exciting project of his career. The *Birds of Australia* gave its creator a new lease of life. By a fortunate coincidence, Gilbert had returned to England barely a month after Eliza's death and drowned the sorrows of his employer in a deluge of new specimens. Gould retracted further into his own private and magical world, cocooned from the realities of life without Eliza by his work. In the same letter in which he tells Jardine that 'the loss I have sustained is indeed the greatest that could have befallen me', he enthuses over the booty brought back by Gilbert from Port Essington: 'Of Quadrupeds among other novelties he has brought two species of Kangaroos quite new, one of them a very noble specimen. Among Birds two species of true Megapodei! whose habits and nidification are most extraordinary, new Chlamydera, Pitta, parrots. His arrival at this moment has occupied me much and will do for some time.'

Eliza's place on the production line was going to prove difficult to fill, but the problem was, in Gould's mind, far from insurmountable. 'The loss of my very efficient helpmate', he told Jardine, 'will necessarily involve me in considerable trouble with respect to the drawing and although I am happy to say I have artists in training who are fully competent for everything that can be wished they require from me more perfect sketches, constant supervision while each drawing is in progress, etc.' Less than a fortnight later, however, Gould was feeling positive and confident about the task in hand: 'You will be pleased to hear', he informed Strickland on 13 October, 'that I am getting my birds beautifully drawn by a talented young artist and as I have ever made and shall continue to make the designs, no difference will be perceptible in the succeeding plates of the work from those already issued.'

The young artist Gould had found was Henry Constantine Richter, son of an eminent historical painter Henry James Richter, member of the Old Water-Colour Society and President of the Associated Artists in Watercolours from 1811-1812. Henry did not quite live up to the artistic prowess of his father, or indeed of his sister, a portraitist currently exhibiting miniatures at the Royal Academy; his talents lay in the less illustrious, and less remunerative, line of illustration.

It was probably the three plates Henry submitted in the previous year for the *Genera of Birds* by George Robert Gray of the British Museum that first drew Gould's attention to the young man's potential as an ornithological draughtsman. Richter found his early days working for the inexhaustible ornithologist taxing. He not only had to accustom himself to the pace and motion of the relentless Gould machine, but he had to compete with the shining example of the late Mrs Gould. The shorthand that had developed between husband and wife over 11 years' collaboration had to be learned by Richter in a matter of months. He not only had to translate Gould's hasty and heavy-handed sketches into beautiful compositions, and to interpret Gould's comments scribbled against his drawings; he was asked to draw strange birds and plants he had never seen alive. It is not surprising that Richter required assistance beyond which Gould was accustomed to give. 'I shall soon be more at liberty', wrote Gould to Jardine in August 1842, 'although my friend Richter always requires me at his elbow and as it were compels me to stay at home.'

Ultimately the new partnership was a success. Over the next seven years Richter completed the remaining plates for *Birds of Australia*, and went on to contribute to subsequent Gould publications, in all some 500 folio plates. Richter's style, self-consciously perhaps, is very similar to Mrs Gould's. His plates are technically proficient, anatomically accurate, and finely detailed; they show a mature and imaginative sense of composition, but, like those of his predecessor, they rarely approach the vigour and expression of Lear. They are the work of a draughtsman, not an artist; they do not contribute anything above the essential qualities portrayed in Gould's original sketches. As illustrations, though, Richter's work was more than pleasing, and there were no complaints from Gould's subscribers.

Gould was still haunted by his memories of the Australian

bush; talking to Gilbert and working daily on the birds he had found there served to heighten a sense of nostalgia for the place. Gould was convinced he would see the continent again, and now that he had found a replacement for Mrs Gould, that far-off hope began to seem a possibility. Shortly before Gilbert left London for his second trip, Gould told Lord Derby, 'I myself look forward to the pleasure of again visiting my adopted Country (that is, if I am blessed with a continuance of the Health I now enjoy) & to carrying on another expedition similar to the one which has been so truly successful.' A few months later he wrote to Gilbert in Australia, 'By the bye I dreamed last night that I was among the Menuras in NSW. Were my publication further advanced I should be induced to take another trip and meet you in Sydney and you must not be surprised to see me there before your return.'

The demands of his publications, however, never presented Gould with the right moment for leaving. The *Birds of Australia* promised to be an even greater task than he had expected; his *Monograph of Kangaroos* was hanging in the balance; and there was still the *Icones Avium* in hand. It was also not long before another project fell into his lap.

In July 1842, after a five-year voyage, HMS *Sulphur* returned to England from her exploration of the islands and coasts of the Pacific. She returned with a significant collection of specimens. Gould was invited to illustrate the bird part of the collection, just as he had for HMS *Beagle*, in a royal-quarto-size publication superintended by Richard B. Hinds, the ship's surgeon. Gould's artist was Benjamin Waterhouse Hawkins. Of the 16 birds represented, Gould was held responsible for naming ten. The first part of Gould's contribution to the *Zoology of the Voyage of HMS Sulphur* appeared in October 1843; the second in January 1844.

Shortly after HMS *Sulphur*'s return, Gould had decided to turn his attention to another rather unexpected field of endeavour. On 2 December 1842 he wrote to Jardine: 'Having many years paid considerable attention to and in fact always been much interested in the South American partridges 'Ortyx' I am preparing a Monograph of this group with figures with the intention of publishing it. If you could give me the loan of Macroura or of any ['other' (crossed out)] species from the West Indies you think I do not possess I should feel obliged.' The following year Gould was back among the museums of Europe with this particular publication in mind. 'Mr Gould is on the Continent', wrote Prince to Jardine on 8 August 1843, 'and writes me word he has gained much information respecting the Orticidae [sic].' He was surprisingly successful, and on 10 October, Gould wrote jubilantly to Jardine: 'During my last Continental trip I visited Paris, Brussels, Frankfurt, Nuremburg, Leipsig, Berlin, Hanover, Amsterdam and Leyden . . . I found many new Ortyx's [sic] nearly the whole of which I was allowed to bring with me to London and I consequently have a surprisingly fine collection, I shall therefore at once set about my proposed monograph.'

Once again Gould was caught up in the momentum of publishing and unable and unwilling to break free. By 1844 he had a considerable number of works in progress, as he described to Lord Derby, one of his most ardent supporters and subscriber to all his works:

'. . . With respect to my present work on "the Birds of Australia["], I must ask the great favour of your Lordships yet waiting a year or two longer before you attempt to bind it . . . I have yet an immense number of Novelties to add, & I hope to render it by far the finest work I have yet attempted.

. . . as regards the Kangaroos I must state that until Gilbert gets round to the East & North Coasts, (which I hope he will accomplish in another Year,) this work cannot be brought to a Close. I shall certainly complete it the moment I can; I must therefore beg your Lordship to keep the two parts you have unbound, until you receive the next with Titles, &c.

. . . the "Icones Avium" is a work which will always be in progress. A Title Page for a Volume will be given when 80 or 100 plates have been published. This Work is intended to make known any new & interesting Specimens, particularly any appertaining to those subjects upon which I have previously written. The third part is nearly ready, & will contain nine new Species of Toucans. [This part was never published, but the nine new Toucan species it contained figured in a second edition of Gould's *Monograph of Toucans*, published in 1852]. The Ortyzida will be published the first moment I have to spare. The preparation of the last-mentioned Work will involve an expence [sic] of from 5 to 7 hundred Pounds before it appears in public. It will, however, appear at no lengthened period, I hope, before the Year is out.'

Part One of *A Monograph of Odontophorinae, or Partridges of America* was published later in October 1844. 'Surely my Lord', complained Gould not without some satisfaction, 'I shall have to avoid the error of getting too many irons in the fire.'

Thoughts of returning to Australia were now banished from Gould's mind by a crowd of new endeavours. With the death of Gilbert in June 1845, and the suicide of his brother-in-law in Sydney in September of that same year, Gould must have felt the era well and truly past. Although work continued on the *Birds of Australia* until its close at the end of 1848, it was now virtually under wraps.

In 1841 the British Museum had made the first purchase of Gould's Australian birds, containing 332 specimens from South Australia. In 1847 Gould decided to offer the remainder to the trustees of the British Museum for a sum of £1,000. It contained all the known species of Australian birds bar five, together with the nests and eggs of most of them. They were '1800 specimens in all, in various stages of plumage, each carefully labelled, with the scientific name and the name of the place where killed.' To Gould's horror the trustees turned him down; he was outraged. Determined to humiliate the Museum for its refusal, Gould decided to sell out to Dr Thomas B. Wilson of the Museum of Natural Sciences in Philadelphia. Dr Wilson had purchased in 1846 the gigantic Rivoli collection, one of the greatest private collections in Europe, comprising at least 12,500 specimens, and was clearly keen to establish a Museum of no mean proportions on the other side of the Atlantic.

On 8 May 1847 Gould wrote with his proposal to Dr Wilson's brother Edward, who was acting on his behalf.

Dear Sir,

You know that we have more than once spoken about the destination of my collection of Australian Birds and you are aware that it was my wish that they should be deposited in a national museum. This, however, I feel will not be the case. I have offered them to the Trustees who now have decided against the purchase; in consequence, I believe, of the Government not being willing to sanction any heavy outlays for such purposes at this moment. Having done what I consider my duty by offering them to the Trustees, I am now in a position to dispose of them elsewhere and I give you the preference. The enclosed is a copy of the letter I addressed to the Trustees [this letter is unfortunately missing] which will serve for your guidance. Since I saw you I have determined upon publishing a work on the Oology of Australia, consequently the eggs, unless you wish it, need not be taken. At a word then the collection of Australian Birds is offered to you without the Eggs for £800. – or with the Eggs for £1,000. – I am sure you do not know or can have taken into consideration the value of this collection. Pray, recollect it comprises the complete Ornithology so far as discovered of one entire quarter of the Globe. The specimens are in the finest condition and their value is certainly highly enhanced by their sexes having been ascertained by dissection and will ever be of interest as being the original of my work. You have already purchased a magnificent collection in Paris. Will you not make it more celebrated by adding this thereto.'

Wilson pounced on the opportunity and purchased Gould's eggs as well as his birds; consequently Gould's *Oology* was never produced. For some the loss of a great national treasure that contained all the birds illustrated in the *Birds of Australia* was too much to swallow without protest. Jardine in particular made a last-ditch attempt to reverse the agreement, although it is not known what plans he had in mind to repossess the collection. He wrote to 'J' Wilson Esq with his plea:

> Jardine Hall by Lockerbie
> 9th July, 1847

Dear Sir,

From the very short time I had the pleasure of seeing you in London and from my being hurriedly called to Scotland I had no opportunity of learning your views as to Mr. Gould's birds, which we (ornithologists) all so much regret may soon leave Europe. I know that you agree with me in thinking that such a collection as that you have purchased as the types of an extensive and important work should not leave this country, but should have been taken by some public museum where they could at all times be accessible for study or reference. Our Government and the trustees of the British Museum did wrong in refusing Mr. Gould's very liberal offer, at the same time I think Mr. Gould scarcely made his conditions sufficiently public.

We cannot however go back to these circumstances, the

plain question to be asked of you is whether any induce-
ment could be held out, which would prevail on you to
relinquish your bargain with Mr. Gould and allow his
collections still to be purchased by this country, or, if there
are duplicates that a series of named specimens authen-
ticated by Mr. G. with his figures should be selected from
the collection. I know that it would be a great stretch of
liberality in you to give up such a collection. At the same
time, under the circumstances you will forgive such an
application being made to you. I will state at once on what
terms if any, you would be willing to renounce your
arrangement with Mr. Gould . . .

> Believe me,
> Sincerely yours,
> Wm. Jardine.

Mr. Gould is aware that I have written to you and on
what subjects.

Dr Wilson, however, was unwilling to cede his fortunate and
unexpected coup, and Gould's collection was shipped directly
to Verreaux Frères in Paris for mounting, and then to
Philadelphia where it arrived in June 1849.

The fate of Gould's Australian birds was felt most acutely by
Dr R Bowdler Sharpe, keeper of the bird collection at the
British Museum, author of the Museum's gigantic *Catalogue of
Birds*, and Gould's protégé towards the end of the ornithol-
ogist's life. In 1893, 12 years after Gould's death he wrote:

'It is not for the present writer, who of all men, has felt
most keenly the absence of the Gould collection, with its
hundreds of types, from the series of bird skins in the
British Museum, to criticise the action of the officers of
the British Museum at this distant date, but there can be
no doubt that, scientifically, the loss of this historical series
was nothing less than a national disaster and one which

unfortunately was irreparable . . . Gould told me in later years that he had never intended that the collection should leave the country and he regretted ever afterwards that, in a moment of chagrin at the unexpected refusal of his offer to the nation, he accepted Dr. Wilson's offer and allowed his treasure to go to America. It is pleasing how, after Gould's death, the Trustees of the British Museum promptly secured his remaining collections of birds and eggs on the recommendation of Dr. Gunter.'

Gould had taught the great scientific institution a lesson it would never forget.

CHAPTER 12

All That Glitters

OF ALL THE BIRDS that landed on John Gould's desk in Broad Street, 'sometimes in packing cases, sometimes in a letter', there was one family to which he was particularly devoted. The miniscule and mysterious humming-bird became Gould's lifelong obsession. By the time of his death he had amassed a collection of 1,500 mounted and 3,800 unmounted specimens. Gould's great, five-volume *Monograph of the Trochilidae, or Family of Humming-birds*, a 'magnificent work' which he began in 1849, is considered by many to be the culmination of Gould's genius as ornithologist and publisher.

Gould was haunted by images of the little birds day and night. The tiny skins that arrived in Broad Street from collectors in Quito or Rio, 'defended by a piece of brown paper and made into the form of a letter', captured the ornithologist's imagination. Inspired by the descriptions of naturalists who had seen the little birds in the Amazonian jungles and the Andean mountains, Gould dreamed of his 'hummers' fluttering at flowers and darting from tree to tree, although he was not to see a single live specimen until he visited America expressly for that purpose in 1857. While Gould's interest in other bird families grew jaded by familiarity and the passage of time, the humming-bird remained as seductive and enticing as the day it first entered his life. He wrote:

> 'That our enthusiasm and excitement with regard to most things become lessened, if not deadened, by time, particularly when we have acquired what we vainly consider a complete knowledge of the subject, is, I fear, too often the case with most of us; not so, however, I believe with those who take up the study of the family of Humming Birds. Certainly I can affirm that such is not the case with myself; for the pleasure which I experience on seeing a Humming Bird is as great at the present moment as when first I saw one. During the first 20 years of my acquaintance with these wonderful works of creation my thoughts were often directed to them in the day, and my night dreams have not unfrequently carried me to their native forests in the distant country of America.'

At the outset of his career Gould had been seduced by the first specimen he saw. In a rare burst of emotion, he waxes lyrical at the memory of this encounter in his introduction to the *Trochilidae*: 'That early impressions of the mind,' he says, 'are vividly retained, while events of the day flit from our memory, must have been experienced by everyone. How vivid, then, is my recollection of the first Humming Bird which met my admiring gaze! With what delight did I examine its tiny body and feast my eyes on its glittering plumage!'

The humming-bird was late in making its début on the ornithological scene. In 1758, when the tenth edition of Linnaeus' *Systema Naturae* was published, only 18 species were known. By the time Gould arrived at the Zoological Society as curator and preserver, interest was accumulating in the bird, and new specimens were flooding in to England from

211

recently explored regions of South America and the Antilles. In 1829 Lesson began publishing the first great illustrated work exclusively on 'hummers', and in the work showed and described about 110 species, many of which were new. At the Zoological Society, Vigors and Jardine were particularly interested in the humming-bird, the latter publishing a *Monograph of Trochilidae* in 1833, with illustrations by the Edinburgh publisher, William Lizars.

The greatest collection at that time lay in the hands of Mr George Loddiges, author of the *Botanical Cabinet*, who had obtained nearly 200 species comprising a collection that was undoubtedly 'the first in Europe'. It was Mr Loddiges whom Gould began to woo at an early stage in his career, and who soon granted Gould unlimited access to his prized collection. The two humming-bird fanatics struck up a friendship that Gould liked to suggest transcended the realms of common experience. 'This gentleman and myself', Gould explains, 'were imbued with a kindred spirit in the love we both entertained for this family of living gems. To describe the feeling which animated us with regard to them is impossible. It can, in fact, only be realised by those who have made natural history a study, and who pursue the investigation of its charming mysteries with ardor and delight.'

As early as 1830, in his third year at the Zoological Museum, Gould paid his friend the greatest tribute science could bestow: at the Society's meeting on 14 December, Vigors announced in the minutes that Gould had named a new species of humming-bird *Trochilus Loddigesii*. A year later, at a meeting of the Zoological Society on 10 January 1832, Loddiges returned the compliment by naming another new species of humming-bird *Lesbia Gouldi*. With unnecessary affectation, considering his lack of schooling in the classics, Gould received the news saying he 'would prefer the designation of *Sylphia*' because its 'motion's light and sylph-like'. But Gould did not argue in favour of his preferred name; he had after all never seen a humming-bird on the wing, let alone a live example of the species in question.

Monopoly of the humming-bird as a subject for publication lay in the hands of the self-acclaimed trochilidist. Apart from the ethical considerations of staking a claim to someone else's territory, anyone interested in publishing a serious work on the species would need to gain access to Loddiges' specimens. All Gould could do was to make sure, should the opportunity arise, that he was in a position to be given first refusal. It is not known whether Loddiges was planning an illustrated work on his humming-birds with or without Gould's help, but when the former died unexpectedly in 1845, Gould wasted no time setting about his own publication, assuming command of the humming-bird domain with ease.

By 1846 Gould had begun work in earnest, stuffing and arranging a rival collection. 'I have now been confined to the house more than three months,' he wrote to William Jardine, '. . . all the time engaged with the Trochilidae and have now a collection worth seeing . . . Loddiges' collection will be kept in the family.' A few months later Prince reported to Jardine that Gould 'has visited France, Germany and the finest parts of Switzerland has enjoyed himself much . . . and brought [back] with him . . . large accessions to the group of Trochilidae of great beauty and interest.'

Gould had plenty of scope for his trophy-hunting on the Continent. Demand for the mysterious little birds was escalating in all the major cities, and the market was flooded with hummers of every description. It would not be long before many species would be on the verge of extinction. The fashion industry had not yet realised the infinite possibilities of humming-birds as accessories although it had already got its claws into the humming-bird hat trade. Most collectors could still justify the booming bird-trade for supplying science with unknown species and awakening popular interest to ornithology. The humming-bird trade in particular became a lucrative business for entrepreneurs in Europe and in South America, as Gould pointed out:

'Both Frenchmen and Belgians have proceeded to South America to procure supplies of these birds, and dealers from those countries have established themselves in some of the cities of that part of the world for the like purpose. From Sta. Fe de Bogota alone many thousands of skins are annually sent to London and Paris, and sold as ornaments for the drawing-room and for scientific purposes. The Indians readily learn the art of skinning and preserving, and, as a certain amount of emolument attends the collecting of these objects, they often traverse great distances to procure them; districts more than a hundred miles on either side of Bogota are strictly searched; and hence it is from these places alone we receive not less than seventy species of these birds. In like manner the residents of many parts of Brazil employ their slaves in collecting, skinning and preserving them for European markets, and many thousands are annually sent from Rio de Janeiro, Bahia, and Pernambuco. They also supply the inmates of the convents with many of the more richly coloured species for the manufacture of artificial feather-flowers.'

With characteristic matter-of-factness, Gould dispels the myths that still prevailed at the time about the humming-bird. Extraordinary creatures they may be, but that did not allow the distortion of scientific facts. Romanticism was constantly condemned by Gould as the enemy of science and the purveyor of ignorance and superstition; he shot it down as accurately as he would shoot a specimen for his own examining table:

'Many really absurd statements have been made as to the means by which these birds are obtained for our cabinets. It is most frequently asserted that they are shot with water

or with sand. Now, so far as I am aware, these devices are
never resorted to, but they are usually procured in the
usual way, with Nos. 10 and 11 shot. Those being the sizes
best suited for the purpose. If smaller shot be used the
plumage is very frequently so cut and damaged that the
specimen is rendered of little or no value. By far the greater
number fall to the clay ball of the blowpipe, which the
Indians, and in some instances even Europeans, use with
perfect certainty of aim . . . In Brazil very fine nets are
employed for this purpose, but how this engine is
employed I am unable to state. Unfortunately for me many
specimens of the fine species Cometes sparganurus in my
possession have been obtained by means of birdlime, and
this is evidently the way in which these birds are captured
in the neighbourhood of Chuquisaca.'

For this reason Gould found it better to dispatch his own
collectors into the field than to rely on the indiscriminate
techniques of natives. Equipped with the knowledge of which
species were likely to be uncommon, rare, or even new to
Europe, field naturalists were scouring the continent for little
treasures. They could be exhorted, at a price, to search for a
specimen of a particular sex or age; or simply for further
examples of a specific species to complete a display.

In a typical letter addressed 'To whom it may concern',
Gould would give precise and exacting details of his wishes. In
1869, for example, he was anxious to investigate the Bahama
islands for new species of hummer. Enclosed in an introduction
to his new work on *The Trochilidae, or Family of Humming-
birds* sent to the Governor of Nassau was a letter requesting
specimens from, in particular, Long Island, Great Inagua and
New Providence:

'If specimens cannot be obtained from each of the above
places or from any other islands in the Bahama group [Mr.

Gould] must be content with and will be most thankful for, examples from around Nassau; still Mr. Gould trusts, that as his object is the furtherance of science, the Governor of Nassau, James Walker, Esquire C.B., will use his influence to obtain for him what he so much desires. Mr. Gould would send out some small (No. 10) shot to shoot these birds with if none is to be obtained in Nassau and the bird cannot be obtained any other way. He would also, if necessary forward a small cake of arseniated soap to preserve them with. The great point is to have the specimen *properly* skinned. Mr. Gould says *properly* because herein lies the difficulty, a difficulty far greater than obtaining the birds. Surely some person able to do this may be found among the Medical officers of the islands or the young Surgeons of the Ships visiting it, if not then he would like to receive them in two other ways: First, carefully pass a little cotton wool dipped in arseniated soap, cayenne or common pepper, or powered arsenic, down the throat, and insert a similar pledget into the abdomen after parting the feathers near the vent and making an opening into it with a pen knife, this being done hang them up in a room *by the bill* and let them dry for a few days; the second mode is to place the bird entire, immediately after it is killed in a small stoppered bottle filled with spirits of wine or any other colourless spirit; neither of these modes however, answer so well as skinning. In nearly every instance the specimens are spoiled by being handled immediately after they are shot, that is before they are cold and the blood from the shot wound has coagulated; a little patience and taking up the bird by the tip of the bill is all that is necessary to obviate this; of course when the skinning process is resorted to the beautiful throat mark of the male should be smoothed down with the point of a large pin or some other fine instrument and kept as unsullied and perfect as possible.

If they can be procured Mr. Gould would like to receive six or seven males and two or three females killed at that period of the year when the birds are in their finest plumage which he believes to be the first four months of the year.

Mr. Gould will gladly pay any reasonable amount for the collection of these birds, either in the Bahamas or to any person in England.'

At one time Gould was offering as much as £50 for another specimen of the fantastic *Loddigesia mirabilis*, originally found in Peru in the 1840s and a prized piece of Loddiges' collection. Despite his efforts, however, a second specimen was not discovered until 1880.

In general terms, though, Gould's exploration of South America by proxy was pre-eminently successful. William Loyd Baily, a humming-bird enthusiast based in Philadelphia, wrote in the introduction to his unsuccessful *Illustrations of Trochilidae* that 'At the present time specimens have been procured and sent to Europe of upwards of three hundred and twenty distinct and well-defined species. This result is in great measure owing to the energetic exertions of John Gould of London whose collectors have distributed themselves throughout the continents of America in search of new varieties.'

Gould's increased activity in the field of the humming-bird stimulated the interests of the competition. In France, Bourcier, Mulsant, Boissoneau, De Lattre, Gervais, and Longuemare, among others, began producing papers on newly discovered species of the bird. Everyone in the ornithological world seemed to be racing to claim the glory of naming a new hummer. As Gould began publishing the successive parts of his *Monograph of Trochilidae* in 1849, others began to contest his ground. Both Bonaparte in his *Conspectus* and *Reichenbach* in his *Systems* succeeded in confusing the humming-bird issue by bringing out their own devisive treatments of the subject. The situation gave rise to what Dr Coues in his *Ornithological Bibliography* graphically described as that 'malignant epidemic which we may call the genus-itch'. It was the completion of Gould's 'splendid monument' to the *Trochilidae* to which Coues attributes the end of this period of excessive accumulation and confusion. Gould's *Monograph* provided the definitive reference work on the humming-bird family; it was the most reliable attempt so far to arrange the species systematically, thereby raising the subject to a scientific level. People might have argued the finer points of Gould's descriptions, but at least the petty squabbles that had muddied the waters of science were now a thing of the past.

Charles Darwin found Gould's comprehensive work on the humming-bird particularly useful. It helped to show him the geographical distribution of the various species, and by indicating the similarities between certain forms it pointed to important evolutionary factors. When Gould sent Darwin a complimentary copy of his introduction to his *Monograph of Humming-birds*, which he had privately printed for his colleagues, Darwin thanked him profusely for the gift, which had 'much interested' him. Once again Gould had provided Darwin with valuable information to support his theories of transmutation:

'I am extremely much obliged to you for your present of the Trochilidae, of which I have read every word (except the synonyms) from your pleasant Introduction to the end. It certainly is a grand use of local distribution; and likewise of diversified adaptation. I was particularly glad to see one sentence, which I shall some day use; on the close alliance of the species in the larger genera I see that you allude to the crossing of birds in a state of nature; I, for one, repudiate this notion. One of ten points which has interested me most (which you will not approve of) is the number of "races" or doubtful species. I think I shall extract all these cases; as it will show those persons who are quite ignorant of Nat. History, that the determination of species is not a simple affair. I congratulate you on the completion of your magnificent work.'

If Gould's *Monograph* satisfied the humming-bird fanatics on scientific grounds, it less easily fulfilled their expectations on the artistic side. The illustrations were simply too attractive, too colourful, and too lavish to serve the purpose of description alone. On every page the humming-birds' tiny bodies hovered in brilliant colours of electric blue, emerald green, bright scarlet; their breasts and heads were glazed with varnish or gilded with gold leaf so that they shimmered under lamp-light. Not exactly the stuff of science. Gould's extravagant humming-birds, however, were not created purely in the interests of science, as he would have us believe; they were intended to become one of Gould's greatest money-spinners, and to focus world-wide upon Gould, rather than merely upon the humming-bird phenomenon. Those who could afford them were, after all, limited. The humming-birds were the perfect *ouvrage de luxe* to tempt the grand and royal ladies of Europe. In a letter to Bonaparte (30 January 1850) he boasts: 'Humbolt has promised to get me the Queen of Prussia as a subscriber to the Humming Birds and I have already obtained the Queen of Saxony, the Princess of Wied and several other noble ladies for whom the work is specially adapted.' He even defied his own professed dislike of the practice of naming species after persons of influence as a means of flattery (he once claimed that, even in the case of eminent men of science, 'I have ever questioned its propriety, and have rarely resorted to it' – although this is patently untrue) by calling one of his most beautiful new humming-birds after the wife of Napoleon III and Queen of France, *Eugenia imperatrix*. Subscribers to his *Monograph of Hummingbirds* included, to Gould's gratification, 'nearly all the crowned heads of Europe'. His conquest of the Continent was now complete.

The regal attraction of the *Monograph* lay partly in the expensive gold-leaf technique Gould used to illustrate the iridescence of the humming-bird' plumage. Gould had made much of this invention, which he claimed exclusively for himself, displaying examples in the Fine Art Court of the Crystal Palace for the Great Exhibition of 1851. Under 'exhibitor 247' the official Royal Commission catalogue reads: 'Gould, J., 20 Broad Street, Golden Square – Inventor. A new mode of representing the luminous and metallic colouring of the Trochilidae, or humming birds. The effect is produced by a combination of transparent oil and varnish colours over pure leaf gold laid upon paper, prepared for the purpose.'

Previous attempts had been made to illustrate the metallic reflections of plumage, for example Audebert and Vieillot in their Oiseaux Dorés at the beginning of the nineteenth century. But Gould crowned himself in his preface with the laurels of success: 'Numerous attempts have been made at various times to give something like a representation of the glittering hues with which this group of birds is adorned; but all had ended in disappointment and the subject seemed so fraught with difficulty that I at first dispaired of its accomplishment. I determined, however, to make the trial, and, after a series of lengthened, troublesome, and costly experiments, I have, I trust, partially, if not completely succeeded.'

Typically, we hear nothing further of the details of these difficulties, nor the trials and experiments by which they were overcome. Gould's colourists – W. Hart, in particular – must have played a part in developing the technique, but, as usual, this is not acknowledged. Bowdler Sharpe recorded that 'Mr Hart commenced working for Mr Gould in the summer of 1851, making the patterns for the Humming-birds and colouring the metallic parts of the bird.' He was to go on to learn lithography and to work as Gould's artist on *Birds of Great*

Britain, the second edition of *Monograph of Trogons*, *Birds of New Guinea* and the *Birds of Asia*, becoming one of the most accomplished, if not the best, of Gould's lithographic artists. Again though a veil of secrecy is drawn over the inner-workings of the great Gould machine, the implications being that all innovations originated from the man himself.

Gould was not impervious to suggestions that he had taken his technique from someone else. William Baily, the humming-bird specialist from Philadelphia, had also used gold-leaf in his illustrations for a humming-bird monograph that was never published; Gould was keen to clear himself of any allegations that he had borrowed from the American as he states in his preface: 'Similar attempts were simultaneously carried on in America by Wm. L. Baily, Esq., who with the utmost kindness and liberality explained his process to me and although I have not adopted it, I must in fairness admit that it is fully as successful as my own.'

The story from the other side of the Atlantic is quite different. Baily's nephew, William Lloyd Baily Junior, in a biographical article published in the 'Proceedings of the Delaware Valley Ornithological Club' in 1919, insists that Gould was entirely indebted to his correspondent in Philadelphia for the technique. 'It has always been represented by different members of the Baily family including his own brothers,' he says, 'that William L. Baily was the first to obtain the iridescent effect on coloured drawings of Hummingbirds, and also that John Gould made use of the process of which Baily had given him exact details.'

Although a friendly relationship always existed between the two men, William Baily Jr was convinced that Gould took unfair, if not dishonest, advantage of his uncle's good nature. His uncle, he explains, was generous to a fault:

'Those who knew him have stated to me that he had a most kindly nature, was polite and considerate to those around

221

him, always interesting, and was the soul of open-
heartedness and unselfishness. If he knew anything or had
anything that he thought would help someone else, it was
his even without asking, and he never thought of reserving
any credit to himself. He was so honest that he expected
others to be honest . . .

He was not only in the habit of sending bird pictures to
his friends and relatives, but of giving away mounted
birds, some of them artistically grouped under glass
cases . . .'

This is borne out by his correspondence with Gould, which he
instigated himself on 31 August 1854 in an attempt to avail the
eminent orni-thologist of some valuable advice:

> 252 Chestnut Street
> Philadelphia 8mo 31st 1854

John Gould –
   Esteemed Friend
I take the liberty of writing a few lines upon the subject of
thy new work on Humming Birds, which through the
kindness of my friend Thos. Wilson of this city, I have had
several opportunities of examining. I have for some years
past been experimenting upon a plan to produce the most
natural representation of the metalic coloring of the
humming bird, but have not until recently succeeded to
entire satisfaction. I have found that in the ordinary way
the coloring matter which is laid on the surface of the
metalic printing, is liable to crack and peal off after being
exposed to the changes from damp to dry weather, and
having noticed what appeared to be a similar effect
produced upon some of the plates in thy work, it occurred
to me to write and suggest a method which I have found as
far as experience yet goes to be quite effectual. It is simply
done by coating the metalic surface with a very thin film of

gelatin and then use colors ground in honey. Pictures made in this way have a brilliant metalic glare, but none of that glassy appearance which in some lights destroys the effect desired, and the color never cracks or scales off.

If at some future time I have the opportunity I should be glad to forward by some of my friends one or two of my plates for thy inspection.

I must also add that the more I examine the work previously alluded to, the more I am struck with its beauty and excellence; and I cannot but feel a friendship and esteem for its author both as an artist and naturalist, and it would give me much pleasure to be a subscriber, did I think my circumstances would admit of it; Should thee approve my suggestion which I submit with all due respect, it would please me much to hear of its success – in the meantime with sentiments of true respect I am thy friend

WM. L Baily–

PS – Nos 4& 5 please me very much – could I obtain a copy of one or both of them without the rest? my address will be found upon the outside of the envelope

Baily's letter was greeted with an inquisitive if dignified reception by Gould, who did not deny the shortcomings of his own technique. With characteristically veiled enthusiasm, Gould invited Baily to give him further instruction:

London, 20 Broad Street
Golden Square
Oct. 19 1854.

My dear Sir,

On my return to London after a seven weeks' absence, I find your friendly letter of the 31st of August and beg you will accept my best thanks for the interest you are pleased to express in my work on the Trochilidae and the suggestion you have made for its improvement. I have not yet had time to ascertain its value, but will take an early opportunity of doing so; in the meantime, if you could carry into effect your kind offer to send me one or two of your plates for inspection, I would be obliged. As you will see by the enclosed, the value of Gelatine as a transparent medium has not escaped my notice, as I have printed on it, but I had not thought of employing it in the say you mention. Any hints therefore as to the mode of application etc, will be acceptable to me . . .

> Reciprocating your good feelings,
> I remain,
> My dear Sir,
> Yours very truly,
> John Gould.

A few months later Baily was obediently dispatching his plates to London for examination by Gould:

> Philadelphia 3rd Mo 31st 1855
>
> Esteemed Friend
>
> I have now the pleasure of waiting upon thee with a few pictures copied from my original plates, which although they possess but little merit, yet they will serve to illustrate the point at issue. I do not place them in competition with any other productions except for the one peculiar quality which they possess, that is their perfect pliability under all the changes in temperature, or moisture of the atmosphere. I have not yet known any of my pictures to crack which have been prepared in this way . . .

Gould would have been able to see immediately from Baily's plates whether the technique was of any use to him, but his reply is equivocal. On the one hand he dismisses Baily's technique as being of no use in the mass-production of plates;

on the other he probes for further details. It is as if Gould wants to elicit the exact details of Baily's method without having to acknowledge that he has received any direct benefit from the information. On 8 August 1855 he wrote to Baily:

> My dear Sir:
>   Your friend Mr. Sharpe duly forwarded me your letter of the 3rd of March and the accompanying drawings, for which I beg you will accept my best thanks. I have examined them with much interest and certainly consider them to be very superior, the Bartail especially being very near to nature. I fear, however, I shall not be able to turn your kind attention to account, my own process being performed with greater rapidity, besides which the making of a single or a few drawings being one thing, the preparation of many thousands another. In the account you have kindly sent me of your modus operandi, you say, "previous to taking my impression, I print the gold or silver foil in its proper place". This passage I do not quite understand; perhaps when you next write, you will be so good as to render your meaning a little more clear . . .'

If Baily responded to this request, his letter is sadly lost. But in April 1856 he was writing again to Gould, enclosing a description and drawing of a new species of hummer just discovered in South America, and mentioning that, 'Since writing last I have produced some pictures much superior to any which I have before executed – in some the metalic colouring when exhibited by gas light glows with the brilliancy of nature.'

Gould's lengthy reply of 4 June 1856, detailing his requests for various humming-birds, including specimens of the one described in Baily's letter, refers only briefly to the question of 'metalic colouring', but he is clearly far from uninterested. 'It is just possible', he says, 'that I may be in America next year, and have the pleasure of seeing you and your improvements in colouring in which I am very glad to hear of your success.'

When Gould did eventually find time for a trip to America the following year, it was Baily who introduced him to his first live hummer in Philadelphia. Gould also states that he 'received many other kind attentions' from his friend during his stay there: did these include a perusal of Baily's perfected colouring technique, of which he was so proud? It would be another four years before Gould completed his *Monograph of Trochilidae*, and it is difficult to conceive that he did not benefit in some way from consulting with Baily and inspecting his drawings in detail while in Philadelphia.

Baily had already suggested to his friend Geoffrey N. Lawrence that Gould was deriving useful information from their correspondence. On 10 November 1855 Lawrence wrote to Baily saying: 'The remarkable drawing of Humming Birds is very beautiful and remarkably true to nature; the metallic colouring is very fine in effect and different from anything of the kind I have heretofore seen. If Gould has availed himself of your instructions in using these tints, he certainly should acknowledge it in his work.'

Gould's acknowledgements, or lack of them, were published in September 1861, when the *Monograph of Trochilidae* was finally completed. But by then Baily was unavailable for comment; he died in May of that same year at the age of 33. It is frustrating that we cannot know Baily's reaction to Gould's preface, claiming he had received no benefit at all from the technique put forward by the young ornithologist in America. Baily is the only one who could have known for sure if Gould had followed his advice.

Back in England the splendid iridescent technique of Gould's humming-bird monograph received unconditional enthusiasm by the British press as proof of the ornithologist's illustrative genius. As admiration for Gould's illustrations grew into a rapturous frenzy, so the humming-birds themselves became objects of public admiration. Scientists and collectors, travellers and leading figures in London society flocked to Broad Street to see Gould's collection. 'We have had the Prince of Wales and his brother here today', wrote Lizzie to her sister Louisa in May 1856, '. . . They were principally delighted with the birds Papa shewed them in the office and which they could handle. They took away with them two hummingbird's nests and eggs . . . The street was in an uproar.'

Gould's home at 20 Broad Street had never been a house of conventional appearance, but the introduction of a humming-bird collection, and the uninterrupted stream of new arrivals of birds threatened to take over the building. With great restraint the collection was successfully confined to a single room: 'The drawing room where first I remember it was very pretty', recalled Lizzie in her memoirs, 'but as father's collection of hummingbirds grew larger & he mounted fresh cases for want of other room they were collected there until it became almost too full to move about. The housemaid was not allowed in it with broom or duster except on rare occasions, so in time it looked anything but pretty & of course it was not used as a drawing room.'

Gould's collection, as impressive as it may have been to the fashionable members of society and to the museum zoologists, did not hold the same magic for those who had seen the bird in nature. The itinerant field naturalist W. H. Hudson in particular found Gould's assemblage of stuffed specimens, to which he was introduced after Gould had moved to Charlotte Street, which had been artificially arranged on branches or in positions of flight, a pathetic and sickening spectacle, as he wrote in 1917:

'I shall never forget the first sight I had of the late Mr. Gould's collection of humming-birds (now in the National Museum), shown to me by the naturalist himself, who evidently took considerable pride in the work of his hands.

I had just left tropical nature behind me across the Atlantic, and the unexpected meeting with a transcript of it in a dusty room in Bedford Square gave me a distinct shock. Those pellets of dead feathers, which had long ceased to sparkle and shine, stuck with wires – not invisible – over blossoming cloth and tinsel bushes, how melancholy they made me feel!'

According to Hudson, Gould's obsession with humming-birds was nothing more than a magpie-like addiction – the material desire for possession and the avariciousness of the collector, not the selfless appreciation of a true lover of nature. 'He regarded natural history principally as a "science of dead animals – a *necrology*",' Hudson wrote. But Gould's passion for collecting was much to the taste of the Victorian public. His archives of bird-skins brought him lasting acclaim, and his vast array of stuffed humming-bird specimens proved one of the greatest and most prestigious attractions during the year of the Great Exhibition. Gould's name became synonymous with his most precious and favourite bird; before he had ever seen it alive.

CHAPTER 13

Tresses of the Day-Star

THE GREAT EXHIBITION of 1851 was an epoch-making event. Set at the pivotal mid-century mark, the Exhibition was claimed by Prince Albert, its 'great author', to represent a 'living picture of the point of development at which mankind has arrived, and a new starting point from which all nations will be able to direct their future exertions'. To the Queen it was no less than 'that towering example of our greatness and our reign'. It became a legendary monument to the Victorian era, a giant testimonial to the triumphant success of industrial Britain. It was also the first exhibition of its kind held on an international scale, embodying the spirit of a scientific union of Europe. More than 40 countries, from 'Chili' to China, 'comprising almost the whole of the civilised nations of the globe', were invited to exhibit their wares at the event, and to share their knowledge and expertise.

This was its declared intention, but in reality the Exhibition did more to fire up the rivalry of the leading nations than it did to pacify them. It emphasised the competition between the monarchies of Europe and quickened the race for colonial conquest. Despite Queen Victoria's pious sentiment at the opening of the Exhibition that 'this undertaking may conduce to the welfare of my people and to the common interests of the human race . . .' the Crimean War broke out less than three years later under a cloud of friction and dissension.

For some, however, the Great Exhibition was all too obviously the biggest display of propaganda the world had ever seen. Over half the floor space was dedicated to British exhibits, while gigantic displays of her raw materials were lumped intimidatingly outside the Crystal Palace as if to prove Britain's boundless resources to the world.

Obsessed with the domination of the wild, with conquering and civilising the world the Victorians dismantled nature like some gigantic clock invented solely for the amusement and the benefit of mankind. Each display was designed to show the triumph of man over the disorganised world in which he found himself. It was the year in which Mary Shelley, the author of *Frankenstein*, died, but her fantasy lived on: never was there a time when man was so eager to mimic the natural world in artificial form and to re-animate the lifeless.

John Gould's part in the Great Exhibition fulfilled the greatest of Victorian expectations. He brought together under one roof 320 different species of one of the tiniest and mysterious birds in the world. Gould's humming-birds became a living legend. Preserved, stuffed, and mounted under 24 specially constructed glass cases, they paraded their gorgeous colours for the greedy eyes of the public. 'They hang amidst fuchsia flowers, or float over beds of bromelia. They sit in their nests upon two white eggs, ready to disclose their "golden couplets". They dart long beaks into deep, tubular, flowers, hovering beneath the pendant bells. They poise themselves in the air, we hear not the humming of the wings, but we can almost fancy there is a voice in that beauty,' reported Charles Dickens' *Household Words* rapturously, in an article headed 'Tresses of the Day-Star'.

The Formosan jay (*Garrulus taivanus*), drawn in pencil and watercolor and attributed to H. C. Richter, for his lithograph in *The Birds of Asia*, with annotations by Gould.

A pencil-and-watercolor sketch of Langsdorf's aracari (*Pteroglossus langsdorffi*), artist unknown, with pencil notes by Gould, for *A Monograph of the Ramphastidae, or Family of Toucans.*

a

b

c

d

The production process of a plate showing the elegant pitta (*Pitta concinna*),
one of the birds first described by Gould. The stages
(*opposite*) are: a) the rough watercolor sketch, probably by
Gould himself; b) an ink-and-watercolor drawing by William Hart,
annotated by Gould; c) a transfer tracing, incorporating changes;
d) a lithographic print, with penciled instructions; and (*above*)
the finished hand-colored lithograph.

The green woodpecker (*Gecinus viridis*), drawn in pencil and wash and dated "20 May 58." Penciled instructions are written in three different hands: "less black" (Gould); "stand up" (Richter); "bare swolen skin" (unknown). The long instruction on the right in ink is in Gould's hand and is typically brusque—"all parts to be more *neatly* drawn."

Detail of the water rail (*Rallus aquaticus*), drawn in pencil, watercolor, and wash and dated "Jany 16 57." The sketch is attributed to H. C: Richter and is annotated by Gould: "Correct colours of the soft parts of male when newly killed by me at Osberton."

Gould's plate of Darwin's rhea
(*Rhea darwinii*),
the "small ostrich" that was
partially eaten by
Darwin's shipmates
and reconstructed by Gould.

One of Gould's plates of the
large-beaked Galapagos finches
(*Geospiza strenua*),
from Darwin's *The Zoology of the
Voyage of the Beagle.*

The Icelandic gyrfalcon (*Falco islandus*),
one of the outstanding lithographs of birds of prey by
Joseph Wolf for Gould's *The Birds of Great Britain*.

The common heron (*Ardea cinerea*) by John Gould and H. C. Richter,
from *The Birds of Great Britain*. Gould was the first ornithologist to include
chicks in his illustrations—a charming detail that greatly appealed to Victorian sentiment.

The interior (*above*) and the exterior (*below*) of Gould's temporary
hummingbird house in London's Zoological Gardens. The exhibit attracted over
eighty thousand visitors in 1851.

The entire display was a testimony not only to the genius of John Gould, but to the spirited determination of his collectors, to the conquering of mountains and the penetration of the darkest jungles, to the far-reaching arms of the British Empire itself. If the British had no claim to territory in South America, they could at least steal the glory of naming some of its inhabitants. The taking of a new species was, as *Household Words* declared, one of the most prestigious forms of plunder to be had from a foreign country. Only ten species of hummingbird were known to Linnaeus; Gould had collected over 300, many of which were only recently made known to science.

Gould soon realised that to show his collection off to its very best advantage, he would need a venue separated from the bedlam of the Crystal Palace, yet still associated with the great event. What better place than his old stamping ground, the Zoological Gardens in Regent's Park, to which most of the visitors to the Great Exhibition would irresistibly be attracted? With the air of someone bestowing great favours, Gould addressed the members of the Zoological Council with his proposition, taking care to include every consideration for his own financial gain:

Feb. 5: 1851

Gentlemen

Many scientific friends having strongly urged that during the approaching Exhibition the public should be allowed an opportunity of inspecting my fine collection of the Trochilidae or Humming Birds, I am induced to enquire if some arrangement may not be made for its Exhibition in the Gardens of the Zoological Society? The Collection in itself is one of the most complete extant, comprising as it does nearly 2000 specimens of 300 species (with in many cases the nests and eggs) of one of the loveliest groups of birds in existence and which are conspicuous alike for their variety of form and for their wonderful colouring and brilliancy. I feel confident that its Exhibition would form as important a feature as that of any other in London; at least such is the opinion I am induced to form from the numerous applications that are made to me for its inspection, the granting of which forms a most serious impediment to my scientific pursuits. The Collection having cost me a small sum, it would be imprudent to allow of its exhibition without deriving some advantage therefrom; on the other hand I feel certain that if exhibited in the Gardens of a Society with which I have been so long connected it would materially increase their receipts and I consider that I am only doing my duty in endeavouring to make some arrangement with you in the first instance. Your Secretary is confident of its success and with his concurrence I beg to leave to submit to your consideration the two following propositions.

1. That the Society shall erect a house of about 60 feet by 30 (under the superintendence of the Secretary and myself) with suitable approaches wherein to exhibit the collection for 6d each person on Mondays and one shilling on the remaining weekly days; Fellows and two friends personally introduced free and also free on Sundays: The Society to be entitled to half the receipts. or

[2.] I will pay the erection of a house as above a sum not exceeding £700, the house to be the property of the Society at the expiration of six months from the first of May; but in the event of the receipts not equalling the cost of the erection the Society to pay the difference in consideration of the house becoming their property so that no loss may fall upon myself; the admission to be the same as in the foregoing proposition; the receipts being exclusively my own.

In the event of either of these arrangements being acceded [sic] to I should require permission to sell in the room a small popular Catalogue of the Collection, the price of which would probably be 6d each.

    Awaiting your decision

      I beg to remain, Gentlemen, Your very obt. St.

        John Gould

In addition, Gould asked that the collection be insured for a staggering £3,000.

Gould was assured a favourable response mainly because the Prince Consort, mastermind of the Exhibition, had been made President of the Zoological Society that same year. The Council accepted Gould's second alternative, with all the receipts going to Gould.

Had the Society been less cautious it could have received some of the revenues engendered by Gould's enterprise. As it was, it could only profit from association with Gould, who estimated that by 5 November over 80,000 people had visited his exhibit. A three-page accounting ledger, meticulously kept and labelled simply 'Humming Bird Account', shows that for the period 17 May to 8 November 1851, he received a total of £1589.15.1; £463.17.0 was received for the month of July alone.

Gould's humming-bird exhibition was received rapturously by the public, the press, the scientific fraternity, and Queen Victoria. On 10 June the Queen and Prince Albert, accompanied by the Princesses, the Duke and Duchess of Saxe-Coburg, and Duke Ernest of Wûtemberg, visited the gardens of the Zoological Society. 'After breakfast', wrote the Queen in her journal, 'we drove with our 3 girls, Alexandrine, & the 2 Ernests to the Zoological Gardens, where we saw the lions, tigers & leopards, which are very fine, – also a collection (in a room specially arranged for the purpose) of Gould's stuffed Humming Birds. It is the most beautiful & complete collection ever seen, & it is impossible to imagine anything so lovely as these little Humming Birds, their variety, & the extraordinary brilliancy of their colours.'

The appeal of the humming-bird display was in keeping with the character of the other attractions of the Great Exhibition – so much so that the little birds almost lost their identity in the pageant of opulence and ornament. Just as the Queen's gigantic diamond was displayed like a bird in a gilded cage, so were Gould's birds arranged in glass cases like the precious stones to which they were so often compared. Science was partly responsible for the association. The prize of possessing a new and rare specimen was as prestigious in ornithology as it was in geology; the price paid for a single bird-skin could be as much as that offered for a precious stone. By the middle of the century rare birds were regarded as collector's items, to be shown off like a woman parading her jewels. Less than 40 years later the trade in humming-birds reached its peak; millions of the tiny birds were slaughtered for the sake of fashion. In one week in 1888, 400,000 were sold at auction; 12,000 in a single public sale in London on 21 March. The brilliantly coloured ruby and topaz humming-birds suffered especially with 3,000 skins shipped from Brazil in one consignment.

Small wonder, then, that Gould's collection was known as 'the treasure of ornithology'. If Gould could not be credited with actually creating the myriad varieties of hummer, he could at least be marvelled at for his efforts in bringing them together. He was a master showman, and his exhibit was planned as meticulously and ingeniously as that of the best stage-manager. The interior of the humming-bird house was designed to reflect the metallic sheen of the bird's feathers. A number of the same species were displayed in a single case as Gould explicitly directed, to maximise the effect of the same colours together. Great care was paid to the background against which the birds were to be shown.

Inside the building light was theatrically directed to highlight the tiny birds and to play upon their feathers: 'the cases were placed in a large room with light coming from above them, and each bird was carefully set at the angle that would best display the metallic lustre of its plumage, producing separately and collectively a fine effect of colour.' Meanwhile the specimens themselves were 'set upon almost invisible wires' to give the illusion of motion – they were, commented one reviewer in *The Times*, 'tremulous as when during life they hovered over the blossoms of a Mexican wilderness'. The construction of the unique 'revolving cases' themselves, which allowed the viewer to pass all the way around them, produced a three dimensional effect in contrast to the static arrangements of regular table top displays.

The humming-bird exhibition was not only a tribute to the mysteries and wonders of the natural world, and a fanfare to modern scientific man – it was also a masterpiece of public relations. In the very year that his arch-rival John James Audubon died in New York, Gould established himself in London as one of the greatest ornithologists alive. So successful was his exhibition and the reputation it earned him that Gould was reluctant to stem the flow of visitors with the official closing of the Great Exhibition. He hastily made provisions for its continuation, even though this involved another cash investment and the construction of another building:

> 20 Broad Street
> Golden Square
> No.19: 1851

Gentlemen,

It will be remembered that I lately submitted to you some propositions for the continuance of the Exhibition of my Collection of the Trochilidae in the Gardens for 12 months longer and that at your last Council you were pleased unanimously to assent thereto: learning that the Commissioners of Her Majesty's Woods and Forests have determined that the Building in which the Collection is deposited shall be taken down; under these circumstances, feeling confident that the immediate removal of the

Collection from the Gardens will be a disappointment to many of the Fellows as well as to the public at large, I beg to say that I will re-erect the building at my own cost on any other site you may please to select and at once proceed to render the Collection as perfect as possible whereby its importance and attractiveness will be greatly increased.

I have not failed to observe the pleasure which an inspection of the Collection gives to the Fellows and their friends who visit the Gardens on the afternoons of Sundays, I therefore propose, in the event of the Collection remaining in the Gardens for 12 months longer, to allow the fellows to introduce four persons instead of limiting them to the admission of themselves and two friends in their company as hitherto.

Awaiting your decision

I beg to remain, Gentlemen,

Your very obt svt

John Gould.

To some of his subscribers Gould was content to give the impression that the exhibition's extension was a selfless concession to scientific, as well as public, demand. In a letter to R. J. Shuttleworth, dated 9 April 1851, in which he encloses several numbers of his works, Gould says, 'I do not know if you visited my collection of the Trochilidae during its exhibition in the Gardens of the Zool. Soc. if so you will be interested to know that it was so attractive that at the request of the council I have lent it to the Society for the next season and the visitors to the Garden will be admitted without extra charge.'

By then the collection was world famous. The tiny humming-birds had captured the imagination of the Victorian public – Audubon's 'glittering fragments of the rainbow' had nothing on the appellations that Gould's 'feathered jewels' attained. Within the context of the Great Exhibition, Gould's humming-birds were seen as God's own contribution to the event – a blessing conferred by that ultimate inventor, the Creator himself, on the glorious achievements of Britain's Industrial Revolution.

CHAPTER 14

Sir, the House is Unhealthy

WITHIN THREE YEARS of the Great Exhibition a disaster struck London that ripped away the glittering façade of Victorian opulence and exposed the wretched conditions of the workforce upon which it depended. In 1854 a cholera epidemic swept the capital, claiming the lives of 10,738 people. Paxton, the great architect of the Crystal Palace, turned his hand to the less glamorous but more practical task of creating burial grounds. The 'single most terrible outbreak of cholera which ever occurred in this kingdom' originated from a water-pump outside John Gould's house in Broad Street.

Soho had long been the grey area of everyman's land, where the affluent West End met the working-class East End of London. In particular it had received a great influx of people as a result of the Industrial Revolution and the Napoleonic Wars; and in the years leading up to the cholera epidemic mass immigrations of Irish, fleeing from the potato famine in their own country, had arrived: in 1841 there were 75,000 Irish living in London; by 1854 this number had more than doubled. Conditions in the poorer quarters of the capital were appalling, and Soho was one of the most densely populated and insanitary areas of them all.

Side by side with the residential houses owned by single families were squats and slums where, on average, eight people slept in one room. A few yards away from Gould's house in Broad Street, the Poland Street workhouse boasted 500 inmates.

By far the worst problem was sewage. Unlike the privileged Londoners on the other side of Regent's Street, few of the houses in Soho had their own piped water or toilets. The sewage system was ancient and decrepit; overflowing cesspits welled up into basements of the houses over which they were built if there were heavy rains; raw sewage sometimes trickled down the gutters in the open street. Drinking water was collected from various street pumps, which also served as a source for washing water for people in the area. Dickens describes every doorway in Soho being 'blocked up and rendered nearly impassable by a motley collection of children and porter pots of all sizes, from the baby in arms and the half-pint pot, to the full-grown girl and half-gallon can'. The pump in Broad Street as Gould's daughter Lizzie recalled, was a focal point of everyday life:

'One of our great amusements was watching the grooming of [father's horse] Georgie in the yard below from our high nursery window at the back of the house. Another was to watch the commers [sic] and goers to the pump in the Street from the room called the 2nd floor, which was the school room of a morning, when we grew older, & general living room for us children. There was always *something* going on at the pump, the water was considered very good then & people sent jugs to be filled from all around . . . The water cress women would pump on their cresses, & then sit down on the curb, & shake each bunch & re-arrange them in their baskets, & all day long children were playing round the pump, making drinking cups out of their hands, or caps, sending the handle high up & riding down on it & all sorts of tricks.'

Her memories, however, seen no doubt through the rosy-coloured spectacles of old age, were far from the grim reality and misery of the less fortunate people from the tenented buildings around her. A pitiful *cri du coeur* was published in *The Times* in 1849 from some of the poorest people in Soho:

THE EDITOR OF THE TIMES PAPER

Sur,

May we beg and beseach your proteckshion and power. We are Sur, as it may be, livin in a willderniss, so far as the rest of London known's anything of us, or as the rich and great people care about. We live in muck and filthe. We aint go no privez, no dust bins, no drains, no water-splies, and no drain or suer in the hole place. The Suer Company, in Greek St, Soho Square, all great, rich and powerfool men, take no notice watsomedever of our complaints. The stenche of a Gully-hole is disgustin. We all of us suffer, and numbers are ill, and if the Colera comes Lord help us.

Some gentlemens comed yesterday, and we thought they was comishoners from the Suer Company, but they was complaining of the noosance and stenche our lanes and corts was to them in New Oxforde Street. They was much surprized to see the seller of No 12, Carrier St, in our lane, where a child was dyin from fever, and would not believe that Sixty persons sleep in it every night. This here seller you couldent swing a cat in, and the rent is five shilling a week; but theare are greate many sich deare sellers. Sur, we hope you will let us have our cumplaints put into your hinfluenshall paper, and make these landlords of our houses and these comishoners (the friends we spose of the landlords) make our houses decent for Christions to live in.

Preaye Sir com and see us, for we are livin like piggs, and

it aint faire we shoulde be so ill treted.

We are your respeckful servents in Church Lane, Carrier St, and other corts.

Teusday, Juley 3, 1849

If conditions at 20 Broad Street did not equal the misery of the poorest inhabitants of Soho, they were still bad enough to give rise to alarm and misgivings among some members of Gould's household. Prince in particular, who had been charged with the welfare of the family while Mr and Mrs Gould were in Australia, tried to warn Gould of the dangers of living there. In a letter dated 20 March 1839, he wrote to Gould expressing his concern:

'I know you will be anxious to know how we are all after the winter: and on this point I am sorry to say my report must be far from favourable. Mr Mitchell has been seriously ill for the last six weeks or 2 months but I hope will be patched up again for some time longer, both your little ones have also been indisposed particularly Miss Louisa who in fact continues so delicate that I have deemed it my duty to speak to Mr Russell about her being sent into the Country which will be done as soon as he considers the season sufficiently advanced to produce a favourable result. You must not imagine that there is any cause for alarm now though at one time Mr Cox assures me that both Mr Russell and himself did not think it possible she could survive. A decided change for the better has, however taken place and she is so much improved that I conceive a residence in the country is the only thing necessary to establish her health. Little Lizzy is now well. Mrs Coxen has been suffering from her usual attack. Mr Russell says she is much weaker than before but that she will soon be recovered. In addition to this I have myself

241

been much indisposed for the last 4 or 5 weeks from cold and constant pain in the chest indeed the whole household has been so sickly that I do hope you will not attempt to reside here on your return for I am thoroughly convinced that the house itself is positively an unhealthy one, in fact so old and so full of draughts it is that it is next to impossible to escape illness.'

Winter in the capital was a particular hazard to health, the fogs from the Thames and the seasonal low-pressures exacerbating the unhealthy atmosphere of places like Soho, and giving rise to all sorts of bronchial complaints. The following year Prince recorded a similar tale of ill health:

Jan. 15: 1840

Dear Sir:

This letter would have been sent as soon after the close of 1839 as I could have made up the a/cs but I have been laid up in bed for the last 8 days with an inflammation of the chest and lungs and have had to submit to their far from pleasant accompaniments leeches, blisters, pills, draughts, etc. etc. I am now slowly recovering but I am so weak at present that I can but first write: this I hope will be a sufficient reason for the delay. Colds, coughs, and inflammations are quite the fashion with us all . . .

Gould did not take his secretary's advice. In fact he remained at Broad Street for another 20 years, despite the worsening conditions of the neighbourhood and the devastating appearance of cholera. This is not to imply that Gould had taken a harsh or even cruel decision to remain in Soho against all odds, even at the risk of his own children's health. He often expressed concern for them, and took them to the country for a spell of fresh air whenever his overloaded schedule gave him the opportunity. Lizzie described in her journal the relief of escaping the city on one of these trips following a typical summons from her father:

'The cab was at the door, and we were at the station and quickly leaving smoky London behind. Our Father was gone for a few days fishing on the Thames and had arranged to telegraph for us the next day should the weather prove fine. We had been in town all Winter, and were feeling the want of a little country change, so when the laconic message arrived, "Come Down", we did not delay to obey orders.

What a treat it is to see green fields and hedgerows again and railway banks studdied (sic) with wild flowers, and get peeps of the clear blue sky between the clouds . . .'

But these jaunts were few and far between, and the burden of living in an 'unhealthy house' must have been trying, not least for Gould's three daughters, who spent most of the time there. Gould appears to have been so engrossed in his work however that he did not concern himself with the 'trivialities' of health and comfort. The prospect of moving his birds, books, and drawings was hardly a possibility for the middle-aged ornithologist. Quite apart from dismantling the organised chaos of his collections, the disruption of moving house would hardly have been tolerated in a schedule so overworked as his. When the cholera epidemic struck right outside his house it passed Gould by, leaving the ornithologist almost reverently undisturbed. None of his letters contain any reference to one of the greatest tragedies ever to hit London.

The third cholera epidemic to sweep Britain reached London in August 1854. It began in the last days of the month with a few scattered cases around Broad Street. Suddenly, on 31 August the fearful monster broke loose and ran rampant through the streets of the immediate area. On 1 September 70 people died; the next day 127; 300 people died in the following week. One man, John Snow, was determined to find the source of the epidemic, which he believed to be the water pump outside John Gould's house. Snow needed confirmation and he found it, among others, in the testimony of John Gould:

'I inquired of many persons whether they had observed any change in the character of the water, about the time of the outbreak of cholera, and was answered in the negative. I afterwards, however, met with the following important information on this point. Mr. Gould, the eminent ornithologist, lives near the pump in Broad Street, and was in the habit of drinking water. He was out of town at the commencement of the outbreak of cholera, but came home on Saturday morning, 2nd September, and sent for some of the water almost immediately, when he was much surprised to find that it had an offensive smell, although perfectly transparent and fresh from the pump. He did not drink any of it. Mr. Gould's assistant, Mr. Prince, had his attention drawn to the water, and perceived its offensive smell. A servant of Mr Gould who drank the pump water daily, and drank a great deal of it on August 31st, was seized with cholera at an early hour on September 1st. She ultimately recovered.'

This was valuable evidence, and there were further testimonies that pointed unquestionably to the Broad Street pump.

Whether Gould came back to Broad Street on 2 September when the epidemic was at its height, to the aid of his servant who had contracted the disease and to support the household, or whether he simply returned in the regular course of his work, we shall probably never know; but he was certainly one of the few who ventured undaunted on to the scene when all around were rushing for cover. Neither cholera nor any other consideration for health could deflect him from his finds. Isolated in his own impenetrable world at No. 20, he rose above the fear and threats that infiltrated Soho in the wake of the epidemic. The continuation of his work was Gould's principal preoccupation.

By the time of the epidemic Gould was engrossed in five major publications. His regal *Monograph on the Hummingbirds* had reached its seventh of 25 parts, the glittering pages finding their way as regularly as clockwork to the courts of Europe; and *Mammals of Australia*, begun nine years before, was nearing the half-way mark. Gould had also been persuaded by the number of new species still arriving from his 'adopted Country' to embark on a *Supplement to the Birds of Australia*. Gould felt that the plates for this addendum ought to be published in close succession so that the five parts could be bound into a single volume, together with the titles and letterpress. Other commitments, however, prevented this: Gould could barely spare his colourists one second from the pressing engagements to hand; in the end it took 18 years to produce the required copies of the 81-page supplement.

Gould was also in the throes of producing a second edition of his single-volume *Monograph of the Toucans*, containing the nine new species that he had hoped, eight years before, to include in the third part of *Icones Avium*. The latter however never appeared, and with even more new Toucans to hand, Gould was convinced of the validity of publishing again.

His most demanding preoccupation, though, was the mammoth *Birds of Asia*, a publication begun in 1849 and destined to run for 34 years, at the rate of roughly one part (containing 17 plates) a year. Although, as Gould confessed, many species that appeared in the *Birds of Europe* and the *Century* would also appear in the work on Asia, the seven-volume, 530-plate publication would represent the conquest – in the style of an ornithological Marco Polo – of the largest continent of the world. Gould would become the furthest ranging of all ornithologists; he would be a swallow among sparrows, an albatross among gulls. With such ambitions constantly before him, it is perhaps little wonder that the melée in the streets of

Soho could not distract him from the wastes of the Siberian tundra or the rainforests of South America.

Personal tragedy, which came in the wake of the national disaster, also had little impact on the ornithologist's work. Gould's eldest son, John Henry, had left London at the age of 24 on 22 January, seven months before the epidemic, to assume his post as assistant-surgeon in the service of the Honourable East India Company. Gould had accompanied him in the course of his ornithological work as far as Malta, and the two had parted company on 8 February, Gould to return to London, Henry heading for the Red Sea.

Henry wrote his father 12 letters between February and November 1854, describing in lively detail the mosque and bazaars of Cairo, street-fighting in Alexandria, the joss houses and 'Hindoo temples' of Bombay, and his government duties in 'Kurrachee'. While Gould remained aloof from the threat of cholera raging about him, his son, confronted daily during his travels in the Near East with the dreadful realities of the disease, was more concerned. Here, in the very birthplace of cholera, Dr Henry Gould was surrounded by the constant threat of death.

On an overcrowded boat from Bombay to Karachi, Henry told his father, 'Cholera appeared, as might be expected and in six hours killed a man.' When he heard that an epidemic had broken out in Soho he was understandably alarmed. It was left to Prince, however, to allay the young man's fears. In his last letter to his father Henry acknowledged the news:

Camp, Kurrachee, Nov. 7, 1854
Thermometer, morning, 62
12 o'clock, 89

I have just received Mr. Prince's letter, which has relieved my mind of considerable anxiety concerning you all, on account of the prevalence of cholera. I regret much that so many of our neighbours have been swept off by it, and esteem it a great mercy that our family has been spared. At Ghiznee I had always before me an evidence of the dreadful effects of this great modern scourge; for only a few years since, Her Majesty's 86th Regiment lost nearly 400 men in one week by cholera, and their bodies were thrown into large pits close to the site upon which the Sanatarium is now built, and a few mounds are now the only traces of the calamity . . .

Henry's letters to his father make amusing and colourful reading; they also dutifully reflect on the local bird life and the popularity of Gould's publications in the various places he visits. It was probably on both accounts that Gould decided to publish them 'for private circulation' in a small pamphlet, ostensibly, as he put it, 'to answer the following inquiry from many kind friends:– 'Have you heard from your Son lately; how is he getting on, and what is he doing?'

Gould was immensely proud of his eldest son, and of his scientific abilities in particular. As a child in Tasmania, Henry had shown his mother's talent for drawing, and there are a large number of watercolours of eggs signed J. H. Gould, which appear to have been executed in preparation for a work on the *Oology of Australia*, another of his father's publications that never came to fruition. While in the East, Henry had been encouraged by his father to collect bird specimens, 12 of which, although not new, were exhibited by Gould at the Zoological Society on 27 March 1855. Henry also planned to undertake a scientific journey from Karachi to Khelat on behalf of his company. 'Please to let me know how to make "damper"', he asked his father, 'as, if I go, I shall have to be my own baker for a few months; I should like to know this by return of post, and also to receive any other hints your bush experience enables you to give.'

Gould must have felt that in some respects Henry was travelling in his footsteps. The publication of his letters reflected not only the charming character of Gould's son and his scientific inclinations; it also advertised the influence of his father. 'I wish that you could be with me', wrote Henry, 'the trip would then be a great treat, and I might learn something.' In the midst of describing the exotic and barbaric scenes of the East, he told his father dutifully, 'Ornithology, you may be sure, will not be forgotten.'

When Henry died a year later in Bombay of 'a fever', John Gould might have made the same reassuring comment to his public. Distraught and shocked he may have been, but he would not be distracted from his subject. On 8 December 1855, barely two months after the death of his son, Gould wrote with dignified composure about his bereavement to Charles Bonaparte, before getting back to business:

> My dear Prince,
>
> I feel very grateful to your Highness for the kind sympathy you have been pleased to express in the heavy loss I have lately sustained in the death of my eldest son. A greater could not have befallen me and the affliction is equally felt by the whole of my family and I may say I believe by my numerous scientific friends. With God I am enabled to bear it with fortitude aided greatly by knowing that I have allways [sic] done my duty to all my children and by the pleasing occupation in which I am engaged [viz:] the Science of Nat. History in which we are both so much interested.
>
> Having said thus much in reference to the Memory of him I have lost – I now turn to another subject. You will doubtlessly recollect my shewing you a female Humming bird . . .

Gould's obsession became an unusual but merciful blessing. Just as in the case of his wife and the cholera epidemic of 1854, Gould greeted the death of his eldest son with a resolute barrage of inner strength. His work served him once again almost as a religion.

CHAPTER 15

In Quest of a Living Hummer

B Y EARLY SUMMER 1857 Gould had seen his way clear of the publications that lay seige to his office and took a few months leave to realise his life-long dream: to see a live humming-bird. He took with him his second son, 23-year-old Charles, who had recently graduated top of his class from the London School of Mines at South Kensington. The two left Liverpool at 3 PM on Saturday, 2 May on the British Royal Mail Steamship, the *Asia*, bound for New York. They arrived 13 days later in the strange and extraordinary 'empire city' of America.

New York was already showing its colours as the great commercial centre of the Union. It had a population of over a million, and was growing daily due to the influx of immigrants arriving from Europe. The opportunities available to the enterprising 'nobody' were already legendary. Even in its golden days of promise and prosperity, however, the city was tarnished by an underclass of jobless, penniless and hopeless misadventurers. An aura of commercialism hovered above the streets; the cut-and-thrust tussle for a bite of the Big Apple had begun.

Gould was impressed, but more perhaps by his first sight of the city from the sea, which refreshed the eye of the weary traveller and welcomed him to the New World, than by the bustling metropolis. The day after his arrival, Gould wrote home to Broad Street, characteristically keeping his descriptions – despite his obvious interest and enthusiasm – to the barest minimum:

> New York
> 16th May 57
>
> My dearest children
> I take the earliest opportunity after arriving to say we are well, in a day or two either Charles or myself will write more fully. Indeed it is just probable that our second letter may arrive first. The entrance to New York from the sea is truly beautiful and from what I have seen of the place (I mean the city) I am not a little astonished and amused. I trust in a day or two to change this turmoil for quieter scenes in a state of nature. Trusting my dearest children that you are all well and that a kind providence will ever protect you. I will ask you to kindly remember me to Miss Yates, Mr. Prince, and all enquiring friends, and believe me to remain, my dear children,
> Your ever affectionate father,
> John Gould.

There were many places in and around New York City that were hospitable to the tiny humming-bird: the banks of the Hudson River were 'finely wooded'; Central Park offered a great, calm oasis to birds of every variety; and cemeteries such as the Greenwood in Brooklyn, with its fountains and weeping-willows, preserved something like a state of nature in the suburban sprawl. But it was to be another five frustrating days travelling before Gould would see the object of his dreams. 'The period of my visit to America being somewhat early in the season', he wrote, 'my attempts to discover a living "Hummer" in the neighbourhood of New York during the second week of May were futile.' His only encounter with his favourite bird was the brilliant-fronted emerald of the familiar stuffed variety, which he was given by George N. Lawrence of the American Museum of Natural History.

Although disappointed Gould was determined to make the most of his time in the city of opportunity by forging important contacts and testing the ground for possible subscribers – and he made George Lawrence his American agent.

As soon as he could, Gould travelled south to meet the tiny migrant bird from Mexico. In Philadelphia he sought the services of his recent correspondent and rival illustrator William Baily, who took him immediately to the famous Bartram's Gardens, where, at last, Gould's prayers were answered. With an enthusiasm undimmed by the passage of time, Gould registered in his *Monograph of Trochilidae* beside his plate of the ruby-throated humming bird, the sacred moment when the picture of his imaginings finally became flesh. 'It was on 21st of May, 1857, that my earnest day thoughts and not infrequent night-dreams of thirty years were realised by the sight of a Humming Bird.' The tiny creature hovered at the shrubs, its wings vibrating in hazy semicircles at incomprehensible speed. As it danced lightly from one flower to the next, its flight accompanied by a faint humming sound, it seemed more like a giant bumble-bee than a bird: it did not glide through the air with the quick, darting flight of a swallow or swift, but hung tremulously taking nectar before rising perpendicularly, spinning round, or even flying backwards towards the next tree.

Gould was amazed, but did not allow his astonishment to distract him from his observations. To him, he said, the humming-bird's 'actions appeared unlike anything of the kind I had even seen before, and strongly reminded me of a piece of machinery acted upon by a powerful string. I was particularly struck by this peculiarity in the flight, as it was exactly the opposite of what I had expected.' Gould was eight years into his humming-bird publication and had only just learned how his subject flew. 'It was a bold man or an ignorant one', said Alfred Newton, 'who first ventured to depict Humming Birds flying', doubly so, one might add, if he had never seen one alive. Gould's humming-birds had been portrayed in the *Monograph of the Trochilidae* suspended in air, with motionless, out-stretched wings; their bodies, it seems, defying the laws of gravity by hanging on tiny fragile limbs. Even after his visit to America, Gould continued to depict his humming-birds this way, opting for the scientifically useful rather than the visually realistic. 'It cannot be denied', commented Newton, 'that representations of them in that attitude are often of especial use to the ornithologist'; it was clear where Gould's loyalties lie.

While in Philadelphia, Gould took the opportunity of paying a visit to Dr T.B. Wilson at the Academy of Sciences, the man who had taken on Gould's magnificent Australian collection. At the Academy Gould made a pencil-sketch of a bird that is thought to be the yellow-billed cuckoo. Since none of Gould's artists were with him, it is the only drawing that is certain to be by the ornithologist. Beneath the bird, annotated perhaps by Baily, is the comment 'Sketched by Mr John Gould of London at this Academy may 1857.'

Despite his glimpse of the hummer in Bartram's Gardens, Gould's luck in Philadelphia was scarcely better than it had been in New York. The single male ruby throat he had found was only the solitary vanguard of his species, and in order to see more of them, Gould would have to continue his own migration. 'The almost total absence of Humming Birds around Philadelphia', he wrote, 'proved to me that I was still too early for them, the lateness of the season of 1857 having retarded their movement. I therefore determined to proceed further south to Washington, where in the gardens of the Capitol, I had the pleasure of meeting them in great numbers; in lieu of the single individual in Bartram's Gardens, I was gratified by the sight of from fifty to sixty on a single tree.'

Gould must truly have felt he was in the land of his dreams. Humming-birds buzzed in clusters around the great horse chestnut trees, their bodies gleaming in the bright sun of early summer. He decided to take one alive; a captive ruby-throat became the object of the ornithologist's devotions for the four days or so he remained in the capital. It went with him everywhere: on his visits to the Smithsonian to see Spencer Fullerton Baird and Joseph Henry, and, in all probability, on his excursion to meet the American President, James Buchanan, and the new British Ambassador, Lord Napier.

'A Trochilus colubris captured for me by some friends in Washington immediately afterwards partook of some saccharine food that was presented to it, and in 2 hours it pumped the fluid out of a little bottle whenever I offered it; and in this way it lived with me a constant companion for several days, traveling in a little, thin gauzy bag distended by a slender piece of whale bone and suspended to a button of my coat. It was only necessary for me to take the little bottle from my pocket to induce it to thrust its spiny bill through the gauze, protrude its lengthened tongue down the neck of the bottle, and pump up the fluid until it was satiated; it would then retire to the bottom of its little house, preen its wings and tail-feathers, and seem quite content.'

After his brief stay in Washington D.C. John Gould resumed his lightening tour of the States in search of more humming-birds – and, of course, more subscribers – dragging his bemused son dizzily behind him. A letter from Charles to his sister gives a full and anecdotal account of their father's familiar restless energy as the two toured the major cities of the East Coast and made their way to Canada.

June 11, 1/57 Boston

My dear Sarah,

If I remember rightly I am in debt to you on the score of letters, and therefore proceed to wipe off part of it. As you observe from the date etc. we have made the most of our time. In fact we have been dashing over the country at such a rate that I have scarcely had time to think of you all at home, much less to write. In fact my log book from which I expected great things is as yet uncommenced, the books almost unopened, the shell compeller, that extraordinary compound of tin and wire gauze, is rusting from inactivity and the leather fossil bag, after lingering for some time in ignominious obscurity at the bottom of my carpetbag waiting for better days to come, was finally used up as packing material a short time back at Cleveland. Thus much as a sort of apology for not writing more frequently, and next to acknowledge letters No. I, II, III from Mr. Prince with their enclosure from the misses . . .

I dare say you will be wondering what sort of route we have taken, I will tell you.

1. from   *New York* to Philadelphia
          Philadelp – Washington (through Baltimore)
          Washington – Altoona (in the Allegheny
                          Mountains)
          Altoona – Cleveland (on Lake Erie)
          Cleveland – Buffalo
          Buffalo – Niagara. There ought to have been a
                       flourish of trumpets at this
                       point – imagine it–
          Niagara – Toronto
          Toronto – Montreal
          Montreal – Portland (coming through the White
                        Mountains)
          Portland – Boston

I quite forgot what I told you about in my first letter so that I must run the risk of repetition, I believe I said something about N.Y. (The Empire City, or Commercial Emporium) and Phila (The City of brotherly love). At Washington, (city of magnificent distances) pater familias dined with Lord ['Elgin' crossed out] Napier and spent the evening with the President, the small boy was left at home upon each occasion, and consequently adopted for a short time the profession of mud larker, on the banks (the very muddy banks) of the Potomac collecting however Melauria's [sic] in place of halfpennies, and of course humming the appropriate air of "Shells of Ocean".

It was here that we first met with humming birds and coloured gentlemen in any abundance. The first buzzing by dozens round horse chestnut trees in the gardens of the Capitol, the latter especially numerous in and about the hotel. The weather was exceptionally hot, so I found that it was much better to keep them "men and brother" at a distance, and at all events to get to windward of them. N.B. also a great consumption of cobblers. I suppose you will have seen in the Times an account of the row which took place lately in Washington between pluguglys, and some other political body, I forgot its name. This happened a few days after we had left – 6 or 7 men were killed, and 17 wounded – only – much too trifling to cause any excitement here.

From W to Cleveland, from C to Buffalo (rubie H[umming] B[ird]), Buffalo to Ni___ag___a___ra___. I don't like anything sentimental, therefore do not expect raptures and all that sort of thing. Suffice it to say I was *not* disappointed. It is in every respect as vast and glorious as represented. The only humbug about the falls, is the going underneath. I had heard so much about it that I expected this would be attended with some difficulty, and that there would be something to see – all hum – neither the one nor the other.

There being no H B in the neighbourhood, and no savants Mr Gould found the beauty of the falls alone insufficient to attract him more than a few hours, so off we started again before I had hardly got a glimpse of them, which was the more aggravating in as much as he had been saying for the last fortnight, how glad he should be to get to the falls where he might be quiet for two or three days – (Sic vita Gouldi Johnni) – I do not mean to give you even the suspicion of a hint of the sundry bracelets, etc. we bought after a good deal of small talk, with a young lady at a kind of Niagaran edition of the Lowther bazaar.

From Niagara we went partly by rail and partly by boat to Toronto, which is a very large place, and like most American cities very uninteresting. The only novelty about the place, is the use of planks for foot paths and roadways. There is hardly a stone in the place, half the town consists of plank houses. I got a few land and fresh water shells here and had some pleasant days collecting on a kind of sand bank or peninsula running out into the lake. Tell Frank there were no butterflies, no insects of · any kind with the exception of a few very small beetles.

We went to one or two dinner parties and to an evening party at Government house. The hotels here very dirty and wretched.

On reaching Montreal you find quite a different state of things. It is a much older town, and has not those dreadfull [sic] monotonous streets, all straight and crossing one another at right angles. The people are more than half French, though English is mostly spoken. While here we stayed with Mr. Hodges one of the engineers of the Grand Trunk R. R. He was a fellow passenger of ours in the Asia. From Montreal to Portland.

> From P to Boston
> Your affectionate brother
> Charles Gould

In his wake, Gould left a series of natural history societies reeling. In the September edition of the *Canadian Journal of*

Industry, Science and Art, Dr James Bovell of Toronto recorded: 'During the present summer we were visited by Mr John Gould, the distinguished Naturalist, whose chief object in his tour through Canada was for the purpose of studying the habits and manners of the species of *Trochilus* frequenting this portion of the North American Continent.' William D'Urban, Sub-Curator of the Museum of the Montreal History Society wrote on June 29th: 'At the beginning of this month the celebrated Ornithologist Mr Gould F.R.S. . . . . paid Montreal a flying visit of 3 or 4 days . . . He brought his son with him.'

Having toured a total of well over 1,500 miles in a month-and-a-half, father and son left New York bound for home. Not, however, before Gould had found two more live humming-birds to take with him. It would have been a remarkable coup for Gould to bring back the birds alive. Only one previous attempt to transport the tiny creatures across the Atlantic had succeeded. The young humming-birds had survived the arduous voyage to England and were delivered to the devoted, but sadly in the end inadequate, care of a certain Lady Hammond. 'The little creatures', according to Latham, 'readily took honey from the lips of Lady Hammond, and though the one did not live long, the other survived for at least two months from the time of their arrival.' Gould's attempt to bring his favourite birds back to England was an even more pathetic failure. He had only 48 hours back in London in which to revel in the dubious glory of having a live humming-bird in his house.

'The specimens I brought alive to this country were as docile and fearless as a great moth or any other insect would be under similar treatment. The little cage in which they lived was 12 inches long by 7 inches wide and 8 inches high. In this was placed a diminutive twig of a tree, and, suspended to the side, a glass vial which I daily supplied with saccharine matter in the form of sugar or honey and water, with the addition of the yolk of an unboiled egg. Upon this food they appeared to thrive and be happy during the voyage along the seaboard of America and across the Atlantic, until they arrived within the influence of the climate of Europe. Off the western part of Ireland symptoms of drooping unmistakably exhibited themselves; but, although they never fully rallied, I, as before stated, succeeded in bringing one of them alive to London, where it died on the second day after its arrival at my house. The vessel in which I made the passage took a northerly course, which carried us over the banks of Newfoundland, and, although the cold was rather severe during part of the time, the only effect it appeared to have upon my little pets was to induce a kind of torpidity from which, however, they were readily aroused by placing them in the sunshine or in some warm situation, such as before a fire, in the bosom, etc. I do assure my readers that I have seen these birds cold and stiff, and to all appearances dead, and from this state they were readily restored by a little attention and removed into light and heat, when they would "perk up," flutter their little wings, and feast away upon their usual food as if in the best state of health.'

Sadly, even Gould had not fully comprehended that, the adverse conditions of a British climate apart, these tiny ethereal 'spirits of the air' could not be expected to survive on nectar alone. They needed a more practical, if less ambrosial, supplement of protein-rich insects. Without it they would eventually starve. Gould's attempt to describe, depict, and transport these 'rays of the sun' were considered by the naturalist W.H. Hudson doomed to failure. Any attempt to capture a humming-bird, be it in a cage or on paper, was, in Hudson's view foolish, presumptuous, and even sacrilegious.

Gould had come as close as he could to bringing his miracle

before the British public. It was an achievement symbolic of his self-appointed position as an ambassador of science that pre-empted that other tenuous connection between the two continents – the first transatlantic cable, laid with momentary success in 1858 – by bringing his own version of scientific communion home from the States in 1857. When President Buchanan and Queen Victoria exchanged greetings on 5 August 1858 they were both already familiar with the great ornithologist, his passion for humming-birds and his attempt to introduce them to England. The first transatlantic cable also died an early death, failing completely after the famous exchange. The great Atlantic Ocean would hold its sway for a while longer at least.

If Gould's American trip had failed in one respect, it had still fulfilled his wildest dreams, which he felt should be an example to all. 'In passing through this world,' he said, 'I have remarked that when inquirers of a strong will really set themselves to attain a definite object they generally accomplish it; and in my own case the time at length arrived when I was permitted to revel in the delight of seeing the hummingbirds in a state of nature, and to observe their habits in the woods and among the great flowering trees of the United States of America and in Canada.' It was a vision that remained one of the biggest thrills of John Gould's life.

CHAPTER 16

Wolf, Ruskin, and the Birds of Great Britain

I T WAS 1859 before Gould finally moved from 20 Broad Street, where he had lived for nearly 30 years, to a more salubrious neighbourhood at 26 Charlotte Street. It was hardly a dramatic change of address but it had the distinct advantage for Gould of being 'within a stone's throw' of the British Museum's new buildings in Bloomsbury. 'I am living still nearer', he wrote to Dr Hermann Schlegal, 'to the great British depository of Nature's handy works.' Moreover, Gould was now in a position to invite colleagues to stay. Schlegal was one of the first who was assured of a 'pleasant' visit if he came to live with Gould in London for a spell. Gould took pains too to impress upon the young Alfred Newton that his new home was located in a respectable part of town.

> 26 Charlotte St.
> Bedford Square W. C.
> March 22, '61

My dear Sir,
I have just received your note in answer to mine. But

why do you continue to address me at Charlotte St. *Fitzroy* Square, a locality as a young man you doubly know is favoured for women.

Now I have arrived at maturity and am moreover the father of the family. Pray now my dear sir address your future letters to a place where such staid people as myself generally live and not where a hansom's cab man is sure to take me if I do not expressly say Charlotte Street *Bedford Square* at which place I shall be happy to see you when you are in London.

> I am, My dear Sir
> Yours Very Truly
> John Gould

To which Newton responded, teasingly, 'I am sorry for the mistake I made in addressing my letter to you – It is always a grave error when a naturalist blunders about a "habitat" – though I think I have known even celebrated ornithologists do the like.'

The move forced Gould to reorganise his collections and drawings, which gave him, late in his career, the invigoration of a fresh start. It was in 1862 that the long-awaited *Birds of Great Britain* was begun. 'I am now busily engaged', he told Sir Redmund Barry, 'upon a new work on the 'Birds of Great Britain' which I have undertaken at the request of numerous scientific and other friends who are desirous that I should leave behind me a standard work on our native birds. I enter upon the work 'con amore' and as it will probably be the last of my productions you may be assured that I shall spare no pains to render it in every way worthy of the subject . . .'

If the change of address had a refreshing effect on the ageing ornithologist, there was another influence that had the same effect on his publications – the young German zoological painter Joseph Wolf, who had an impact as revitalising and revolutionary as had the young Edward Lear 20 years before.

Wolf introduced the Gould studio to the vitality of nature and dynamic, expressive art. Since the *Birds of Australia* Gould had limited himself to species that could rarely be obtained alive in England, and it was only the advent of the *Birds of Great Britain* series that provided the opportunity of once again taking illustrations from the live bird: Joseph Wolf's passion for 'Life! Life! Life!', as he so emphatically put it, in all of his drawings, whether observed in the wild or resuscitated from a dead skin, injected a new realism into the Gould production line.

It was Gould's remarkable gift for talent-spotting that first brought the young German to England. The 'greatest of all animal painters', as Alfred Newton later described him, came from a tiny village in the remote and beautiful Moselle Valley in the province of Rhenish Prussia. There Wolf spent his childhood rambling amid the great Rhineland forests, studying – and expertly shooting – his favourite products of nature.

Even at this early age Wolf showed a talent for drawing the local birds and animals with which he was so familiar, and he soon excited the interest of the director of the Darmstadt Museum, Dr Kaup, who showed some of his work to John Gould on a visit to London. Gould was so impressed that he gave Wolf his first commission – a charming miniature watercolour entitled 'Partridges dusting'. It was followed by a second commission entitled 'Woodcock seeking shelter', which Gould submitted to the Royal Academy in 1849, where it drew the attention of the animal-painter Edwin Landseer. At the request of the great artist, Wolf's diminutive picture was placed on 'the line' and became an instant success. Woodcock commissions were, in the extemporaneous manner of Victorian vogue, suddenly all the rage. Landseer himself was impressed, and acknowledged 'I have never seen the expression of a bird recorded as Wolf did it.'

Soon Wolf's talents were the talk of scientific circles in Britain. As his striking illustrations for Schegel's *Traitè de*

Fauconnerie – published in Germany in 1844 and the first the artist had ever executed life-size – became known in London, Wolf was acknowledged as an ornithological artist of astounding ability.

Gould's friend Mr Mitchell, the secretary of the Zoological Society, accordingly invited the rising star to London to help him illustrate Gray's *Genera of Birds*. Wolf at first declined, until the traumas and political turmoil of 1848 swayed his decision. Mitchell believed he had 'secured the best available talent in Europe' for the completion of a work he himself was unable to finish, and Wolf's vivid contributions to the *Genera* proved, to Mitchell's own disadvantage, that he was right.

While Mitchell's drawings merely copied, according to the well-versed practice of the day, the 'distortions of the bird-stuffer', Wolf rose above the dissecting table to portray the characteristic expression and attitude of his living subject. 'The great thing I always aimed at', Wolf told his biographer A.H. Palmer (son and biographer of Samuel Palmer) in later years, 'was the expression of Life.' Intimate knowledge of the living subject, its habits and its behaviour, was the key, Wolf believed, to authentic and successful zoological illustration. It was this rare ability to understand, to sense, to feel the nature of an animal or bird that Mrs Gould, Richter, and Hart all lacked. While Gould possessed an uncanny insight into the character and behaviour of birds, he lacked the artistic ability to portray it. Wolf had both the instinct and the art. He was, as he himself remarked, in the awkward if perhaps privileged position of being called an artist by naturalists, and a naturalist by artists. Said Edwin Landseer, quite simply, 'He must have been a bird before he became a man.'

Wolf first met Gould about a fortnight after he arrived in London. The relationship that developed between the two men was as unlikely and as bizarre as Gould's friendship with Lear. By the time the ornithologist, who was then a veteran of about 44, met the artist, a young man of 28, Gould's reputation both as a hard and sometimes unscrupulous business man, and as an egotist with little concern for the pleasantries of good manners, was well known. Armed with this knowledge, as well as his own brand of self-confidence, irreverence for the scientific establishment, and a sardonic sense of humour, Wolf was an equal match for Gould. He had no illusions about the eminent and successful ornithologist. Wolf's depiction of the great man was not without admiration, but it was devoid of the overblown Victorian sentimentality that abounds in contemporary reviews and portraits of Gould. While Gould's 'industry, enthusiasm and perseverance', Wolf says, were 'beyond praise', he 'was a shrewd old fellow', and 'the most uncouth man I ever knew.'

The gulf that was beginning to separate the scientific world from the artistic became clearer to the young Joseph Wolf the more he learned about London life. 'Among the naturalists,' he said, 'there are some who are very keen about scientific correctness, but who have no artistic feeling'. Nowhere was this more apparent than in his relationship with Gould. Despite the polarity of their completely different outlooks, however, Gould invited Wolf, as he had Edward Lear, to visit Norway with him, and Wolf accepted; their single common interest in birds won the day. Gould set off with his unlikely companion and an interpreter in June 1856 on a yacht called, as it happened, *The Calculating Boy*.

For Gould, the purpose of the trip was principally to research the habits of certain migrating species in their summer breeding grounds for his forthcoming *Birds of Great Britain*. 'Desirous like Mr Hewitson to see the Fieldfare in its native woods', wrote Gould, 'I proceeded to Norway, for this and other reasons, in the year 1856, accompanied by Mr Wolf.'

The two naturalists had much to teach other in the way of ornithology, but Wolf was amazed to discover, once in the field, Gould's inability to identify bird-song. Wolf spent hours looking for the nests of the three-toed woodpecker and the red spotted bluethroat by listening to the distinctive calls and responses of the parents to their young. Typically, Gould preferred to attribute this significant discovery, since it was not his own, to luck rather than superior knowledge. 'Mr. Wolf,' he says in *The Birds of Great Britain*, 'who accompanied me to the celebrated Snee Haetten range of mountains, on the 1st of July accidentally discovered some young birds which were just forward enough to hop out of the nest – a great prize to me who had never before seen the bird at this age in a state of nature.'

Wolf's observation may have been a little unfair: these were uncommon birds, and Gould was not familiar with their young, let alone their sounds. The same criticism, however, has been levelled at Gould by the Australian naturalist Allan McEvey, who noticed the distinct lack of aural references in the *Birds of Australia*. Gould does not appear to have been receptive to the strangely beautiful sounds of Australian birds. He was not, unlike Wolf, good at picking up the notes of an unfamiliar species when he heard it in the wild; he was always more visually attuned. But Gould's years of fishing along the Thames, and the occasional holidays in the countryside had gradually made him an authority on local British birds. By the time he was in his dotage, and taking things more slowly, he was an impressive mimic of garden birds.

One of Gould's aims on his Norwegian trip was to obtain a young capercaillie and, having failed to do so, he advertised for it at one of the local farms. As they were about to leave, Wolf related, some labourers – not recognising the great ornithologist – presented Gould with a dead thrush wrapped in paper, and asked for the reward. The trick was so obvious and so clumsy that, for once, the famous wrath of the ornithologist was not invoked. 'Gould threw the Thrush in the spokeman's face with a laugh, in which the men heartily joined.'

Gould was not always such an easy-going travelling companion. Wolf recalled an occasion when Gould made an uncomfortable and unnecessary scene in a farm-house at which they had taken lodgings at Hjerkin. The two naturalists had been given a 'fine large bedroom', and their hosts thought they might be served breakfast in their room. 'Such a suggestion', relates Palmer, 'aroused Gould's dignity (never very sleepy) and the interpreter was summoned in haste. The hostess was solemnly informed that breakfast must be laid elsewhere, and this being immediately done, Gould's anger was appeased. For a time all went well; but present, behind an unnoticed curtain, an old woman burst into a fit of coughing. They had got their breakfast in a separate room – the bedroom of the ancient grandmother of the family. Gould flew into a passion, but could not help laughing at the way he had been tricked.'

The ornithologist, however, was no foreigner to the art of deception, as Lear, Hodgson, Baily and Gilbert experienced in the course of their associations with Gould the publisher. But it was Gould the collector who revealed to Wolf, much to the artist's amazement, the lengths to which he was willing to go to procure yet another precious specimen. Gould was indefatigable in his reconnaissance of the local dealers' shops in London. If he came upon a new bird-skin, said Wolf, 'he would not betray his excitement but would say, "I think I have that; but I wish you would lend it to me to compare." If the dealer agreed, Gould would whisk his prize away to be sketched, often by Wolf, and return it with no cost to himself, having given no indication to the dealer of its value.

Wolf recalled how the ornithologist would arrive in a state of nervous excitement at his house in Berner's Street, just round the corner from Charlotte Street, new skin in hand, and help himself to a cigar, which he would smoke furiously as he passed up and down the room, while waiting for the sketch or watercolour to be finished. On one occasion, the ever-restless Gould decided not to wait while Wolf sketched, but announced he would have his hair cut instead. When he returned, however, he produced with rapture from his handkerchief an egg of the extinct Great Auk. Wolf was astounded and asked where he had found it. 'In the German Bazaar', was Gould's reply,

'Whittaker [a naturalist] asked what he thought a good price, so I gave him a cheque, and here's the egg.' The price was ludicrously cheap. In fact, as another naturalist, Mr Dresser, later elaborated, Gould had already been told about the existence of the egg by the dealer and, having scoffed (perhaps deliberately) at the dealer's description, saying that the egg was in all probability only the double-yolked egg of a different species, went around to the shop to check it out.

Gould's discovery quickly made the rounds, and a collector soon called at his house with an offer for the egg. Gould told him in no uncertain terms that it was the last egg of the species ever likely to come up for sale, and proceeded to ask an enormous sum of money for it. Just as the deal was about to be closed, however, Gould had second, even more profitable, thoughts. 'Wait a bit, sir!', he said to the collector, 'This being probably the last Great Auk's egg which may be forthcoming, I have made up my mind that only a subscriber for one of my works shall have it.' The collector, still longing for his egg but with his arm firmly twisted behind his back, respectfully signed up for the *Birds of Great Britain*, and paid the asking price for the egg as well. 'As for a bargain,' commented Palmer, 'artistic or otherwise, no one could drive a keener.'

Wolf regarded these dealings with amusement mingled with disdain. Even he was not safe from the connivances of his employer: 'I had the skin of a splendid young male Norwegian Falcon,' Wolf said, 'very dark – extremely so. It got into Gould's box, and never found its way out again.'

Gould's admiration for Wolf, however, was unparalleled. He acknowledged Wolf's artistic ability as he never had Lear's. Perhaps it was Wolf's superior field-knowledge that tipped the balance in his favour; perhaps it was simply because Wolf was well recognised in scientific circles, and choosing him as an artist for *Birds of Great Britain* was anything but controversial – in fact it was a selling point. Whatever his reasons, Gould's acknowledgements are profusive and generous. In the text accompanying Wolf's illustration of a greenland falcon, Gould takes an unprecedented opportunity to praise his illustrator:

> 'I cannot conclude without calling attention to the admirable delineation of all the large northern Falcons, for which I am indebted to the pencil of Mr. Wolf, whose abilities as an artist are so justly celebrated, and who thoroughly understands his subject. I trust they will be duly appreciated by the possessors of the present work, as I feel every one must have been delighted with the illustrations of Messrs. Schlegal and Verster van Wulverhorst's 'Traité de Fauconnerie' by the same master hand.'

Wolf indeed had much to offer the Gould production. Gould 'knew little or nothing', the artist considered, about 'designing good attitudes and groups', and in their view it was chiefly for help in composition that Gould took him on board. In matters of style and technique Wolf was also leagues ahead of Gould's workshop illustrators.

Like Lear, Wolf placed special importance on the accurate and sensitive delineation of the surface of a bird – the feathers in particular. Professional naturalists, Wolf believed, rarely had any idea how a bird's feathers behaved in real life: 'To put a bird right,' he said, 'they smooth it down with their own hands.' Wolf claimed to be the first to introduce into England a systematic study of the arrangement of plumage, and that 'the time of its introduction by him may be traced in Gould's works.'

Certainly Wolf's drawing in this respect is outstanding in the *Birds of Great Britain*: a buzzard's feathers are ruffled by the wind; a spotted eagle raises its claw beneath the soft feathers of its chest; the occasional discarded feather rests, as light as air, beside its owner, or is caught in the network of twigs in a nest.

All of Wolf's plates represent a moment of suspended action. Gone are the stilted tableaux of birds frozen in profile for the sake of identification; Wolf's birds all bear the mark of the character of the species. Some, in the style of Landseer, touch an emotive chord: the helplessness of prey and the ruthless power of the victor; the blind greed of young chicks and the sacrificing parent. 'You know', emphasised Wolf, 'I made a distinction between a picture in which there is an idea, and the mere representation of a bird . . .'

Wolf, however, was dissatisfied with his final published work in *Birds of Great Britain*, bemoaning the loss of the freshness and vigour of his drawings once they were translated on to stone. He felt too that Gould, partly through ignorance and partly through sheer bad taste, was attempting to popularise Wolf's illustrations. Palmer leafed through the pages of *Birds of Great Britain* with Wolf to get his reaction. '. . . seeing them [the plates] once more after a long period,' wrote Palmer, 'in the full blaze of their colouring, somewhat disturbed him. He growled over his pipe, as I turned over, such comments as these. Of the Woodcock, "*Much* too red, and he must go and put in those blue-bells and things too! I can't be answerable for the colouring. Everything gets vulgarised." Of the Hoopoe, "Look at that dreadful water he has put in there!" Stella's Duck was "Dreadfully hard, and stripey and streaky".'

Overcolouring was one criticism often levelled at the ornithologist, and the art critic and teacher John Ruskin agreed that Gould lacked many of the artistic qualifications of his chosen profession, as well as being inaccurate in his observations of nature. In his notes on the treatment of the sun and clouds by modern painters, Ruskin points out a fault recurring throughout Gould's fanciful backgrounds for *Birds of Great Britain*:

'The deepening of the cloud is essentially necessary to the redness of the orb. Ordinary observers are continually unaware of this fact, and imagine that a red sun can be darker than the sky around it! Thus Mr Gould, though a professed naturalist, and passing most of his life in the open air, over and over again, in his *British Birds* draws the setting sun dark on the sky!'

Despite exposing Gould's artistic inadequacies, however, Ruskin was keen to stress that the bird illustrations successfully married the aesthetic with the instructive: they were pleasing to the eye, accurately drawn, morally irreproachable, easily understood, and British – the perfect educational tool.

To this end Gould's extravagant taste for colour suited rather than insulted Ruskin's aesthetic sensibilities. His enthusiastic excursions into the paint-pot might have irritated the calm, artistic realism of Wolf, but they played right into the hands of the primary school teacher. With missionary zeal Ruskin set about displaying Gould, Audubon, and other such naturalists in his Sheffield Museum for the purpose of educating the younger generations.

It is clear, however, that in Ruskin's mind at least Gould's birds were confined to the realms of illustration. With Wolf the case was different. He was courted and befriended by Ruskin's beloved Pre-Raphaelites, for whose exhibition in 1857 in Fitzroy Square he contributed several paintings. Although he never quite fell in with the Victorian sentimentalism and anthropomorphism of Landseer, Wolf became widely acknowledged in artistic circles as the German counterpart to England's greatest animal painter.

It is easy to see why Wolf was so readily accepted into the Pre-Raphaelite fold. 'The Pre-Raphaelites imitate no pictures: they paint from nature only', said Ruskin. Much of the Pre-Raphaelite School's inspiration was derived from a new-found affinity with the natural world. No longer frightened by its wildness and immensity – science had seen to that – artists, writers, and poets were ready to restore to Mother Nature the powers of which modern technology had stripped her.

In the spirit of this new sympathy with nature, artists and

scientists began to find the popular fur- and feather-wearing fashions of the day intolerable. Joseph Wolf consistently gave vent to his fury against the contemporary trends of senseless and wasteful exploitation, condemning man as 'the most destructive and carnivorous animal in the world'. Ruskin urged his readers, in his essay on Gould's dabchicks, not 'to confuse God with the Hudson's Bay Company, not to hunt foxes for their brushes instead of their skins, or think the poor little black tails of a Siberian weasel on a judge's shoulders may constitute him therefore a Minos in matters of retributive justice'.

No one recognised the evidence of man's impact upon wild-life more acutely than John Gould. For 50 years he had watched, studied, and assessed the distribution of birds the world over; the devastating impact of the feather trade on their numbers shocked him profoundly. But when the chorus of conversationists began to find their voice in the 1860s Gould's was not among one of the vociferous. He did not have the rebel-liousness of a Charles Waterton, the moral probity of an Alfred Newton or the outspokenness of a William MacGillivray, but he did applaud the bandwagon as it came his way – even if he did not quite jump on it. The anxiety he had felt for certain species in Australia and Van Diemen's Land in 1839 was articulated more forcefully in his *Handbook to the Birds of Australia*, published in 1865, than it ever was in the original publication:

'It may be possible – and, indeed it is most likely – that flocks of Parakeets no longer fly over the houses and chase each other in the streets of Hobart Town or Adelaide, that no longer does the noble Bustard stalk over the flats of the Upper Hunter, nor the Emus feed and breed on the Liverpool plains, as they did at that time; and if this be so, surely the Australians should at once bestir themselves to render protection to these and many other native birds: otherwise very many of them, like the fine Parrot (Nestor productus) of Norfolk Island, will soon become extinct.'

As for British birds, Gould was a keen supporter of the recent legislation that was designed to protect species targeted by the feather trade – as long as it did not infringe on the livelihood of people who traditionally depended on them (provisos already, in fact, contained within the new 1869 Act for the Preservation of Sea Birds).

However, while quoting in his introduction to the *Birds of Great Britain* the impassioned pleas of other ornithologists and bird-lovers who had published articles and letters on the 'wanton, wicked cruelty perpetrated by the ruthless slayers of unoffending birds', and while appearing to agree with their sentiments, Gould's own argument is less forcefully put. Gould's criticisms of the bird trade are to do with declining numbers and breeding rights rather than ignorance, greed or wanton cruelty. Where his public, and most particularly his subscribers, were concerned, Gould as ever sits firmly on the fence. When the main Bird Protection Act came into effect in 1880, Gould was not associated with its instigation.

If Gould was beyond the mainstream of ornithological conservationists, however, his reaction to the feather trade was more expressive in letters to his friends; but here again his grievance is more one of fashion robbing science of its rightful prey and putting up the cost of birds for the honest collector, than the harm this trade was inflicting on the birds themselves. 'The Ladies, the Ladies,' he said in one letter to Frederick McCoy in Australia, in response to the latter's request for specimens for his museum in Melbourne, 'have . . . so stripped us of birds for their bonnets that but few are now in the market and these of course are high priced . . . Birds of Paradise, beyond those you possess, are out of the question . . .' When he acknowledged that certain species were becoming more rare, such as the Dartford warbler, Gould never considered that he should refrain from killing them himself; the concept of individual responsibility was far from the minds of most people in the nineteenth century. Moreover, in the days before photography and film, the gun was the only sure way of capturing a subject, and specimens obtained in the name of science, even of a species on the verge of extinction, were more valuable to the scientist dead than alive. An ornithologist therefore would kill the last remaining example of a species to prove that he had discovered it alive.

This strange state of affairs was well illustrated by ornithologists who considered it legitimate to include in their lists of British birds any bird that had been killed on British soil. Thus a species that had never visited Britain, but had the misfortune to find itself over her shores, would be shot as evidence, and would then be registered by its discoverer as yet another valuable addition to the annals of British birds. Rare migrants and accidental visitors provided an important source of novelty for the ornithological illustrator, including Gould. He was, after all, paid by the plate, not by the volume, and the more birds he could include in his work was to his commercial advantage as well as his prestige.

Gould was not the only culprit: a flock of rare birds needed a miracle to escape the firing squads that bristled all over the country at the hint of a new target. In May 1870, 40 golden orioles were blasted out of the sky shortly after they had arrived in a wood near Penzance. As for the individual interloper blown off course or simply disorientated, there was the chance, provided it could be proved to be 'British killed', of scoring a minor coup. The problems of actually proving that such a bird had been shot in Britain were enormous given the proliferation of dodgy dealers and the tempting incentives for producing that ultimate trophy, the English novelty. In a letter written by Prince and signed by Gould, in which the identity of the addressee is unrecorded, the ornithologist carefully follows up a lead which might, if no one else got wind of it first, provide him with an important British trophy:

London, 26, Charlotte Street
Bedford Square, W.C.
Dec. 14: 1868

Dear Sir,

I am extremely obliged for your note and greatly interested with the news you give me of a great Black Woodpecker having been shot at Benstead: If this be *really true* it is the only authentic instance I can give in my work of the bird having been killed in the British Islands: all previous accounts not having borne the test of investigation: the instances mentioned in our various works rest on statements alone no actually *British killed specimen* being in *existence*. Now I believe *you* would never willingly deceive anyone and especially myself, you will therefore excuse my asking you to *quietly* and without *apparent* interest, ascertain the truth of the bird having been shot at the place mentioned in your note; I ask you to do so for this reason; There is now a system with certain persons, who ought to be ashamed of themselves of getting the rarer British Birds sent from the Continent in the flesh; besides which specimens of them are constantly sent to our markets in the great crates of wild fowl and grouse which come from Norway, Holland, France and even Spain. Such specimens of the Great Sedge Warbler, Squacco Heron, Glossy Ibis, Great Black Woodpecker, Little Bustard and other rare birds with us are at once seized upon by certain conniving persons and forwarded to their correspondents in the country far and wide, some but not all of whom palm them off upon country gentlemen who seeing them in the flesh believe them to be British Killed and readily purchase at enormous prices.

Only three weeks ago a splendid male Black woodpecker was brought to be in the flesh as fresh as a daisy. This bird I ascertained beyond doubt had come direct to Leadenhall market from Norway with some Capercailies [sic] and had been purchased for eight shillings. Now having let you into one of the nefarious tricks of the bird trade, I do not say that the specimen you speak of was not killed at Benstead but I want you for the sake of our favorite science, to ascertain if there has been no collusions. If you are satisfied that there has not, I shall be satisfied also and will figure the species in my work. Go over yourself and quietly learn who shot it, whether a Keeper or a country gentleman, if the bird had been seen before and every particular respecting it; and then let me hear from you again, in the mean while [sic]

    Believe me
        Yours very truly
         John Gould

P.S. This letter is confidential and should not be shewn to anyone.

In the event, Gould's mysterious correspondent was unable to authenticate his bird, and the discovery is not mentioned in the letterpress. Gould decided anyway to include the great black woodpecker in his *Birds of Great Britain*, where its giant body dwarfs the pictures of its more legitimate cousins, despite the fact that none of the alleged sightings of it in Britain were sound. Gould, in fact, figured over 100 birds in his work that were only accidental visitors, 11 of which had only been sighted in Britain once, and one, the Norwegian falcon, never at all.

None of this detracted from the general popularity and acclaim of this patriotic publication. *Birds of Great Britain* attracted 397 subscribers, by far the largest number of applicants for any of Gould's works, and each paid a substantial £78.15.0 for their copy. Today, the complete series is worth in excess of £30,000. The book added significant catches to Gould's list of notable subscribers. With ostentatious satisfaction he notches up the prestigious total in an 1866 prospectus: 'Monarchs, 12 (The Queen and Prince Consort each taking an entire set of the works); Imperial, Serene, and Royal Highnesses, 11; English Dukes and Duchesses, 16; Marquises and Marchionesses, 6; Earls, 30; Counts, Countesses, and Barons, 5; Viscounts, 10; Bishop (Worcester), 1; Lords, 36; Honourables, 31; Baronets, &c, 61; Institutions and Libraries, 107; Miscellaneous, 682.'

The publication was seen – perhaps partly because its subject was British, as the culmination of Gould's genius – more so even than the gigantic eight-volume *Birds of Australia* or the remarkable *Monograph of Hummingbirds*.

Once again, however, the artist who had made such a vital contribution to the work was eclipsed by the status of the author. *Birds of Great Britain* defined, as one article in *Nature* put it, the 'acme' of John Gould's 'artistic talent'. The ornithologist's children were, it must be said, more discerning. 'Take my advice', wrote Charles Gould to his sister Louisa, shortly after their father's death, 'and do not part with one of Wolf's drawings, they will become daily more valuable and the more complete a set you have the more their value will be enhanced.'

CHAPTER 17

The Curtain Falls

I T WAS A DECADE from the publication of the two
textual volumes of the *Handbook to the Birds of Australia*,
published in 1865, before a completely new production
emerged from the Gould stable. Gould was convinced that the
Birds of Great Britain would be his last great opus; it was a
gigantic and all-consuming task, and would, he believed,
provide a fitting grand finale to the monument of work
immortalised in his personal library.

He was, moreover, still engaged in the completion of *Birds of
Asia*, the *Supplement to the Birds of Australia*, and the second
edition of the *Trogons*, and his studio was working at full
capacity. 'I have almost all the colourers in London engaged in
colouring my works in hand,' he told G.N. Lawrence, his agent
in New York, in 1871, 'and more I cannot do.' There was a note
of resignation in this uncharacteristic remark; gradually, the
Gould machine, which had run so smoothly and efficiently
since its construction in 1830, was beginning to show signs of
fatigue. Its days were surely numbered. In 1869 Gould was
suffering his first serious illness. A 'very distressing attack of
blood poisoning' had left him incapacitated for months and,

although he was reviving by the end of the year, the boyish vigour and self-confidence that had characterised the ornithologist throughout the course of his career was forced to change.

Prince too – the linch pin of the company – was falling prey to the pressures of sickness and stress. 'I am sorry to hear also that Mr Prince is again poorly', wrote Franklin to his father from France in November 1869, 'His health is so bound up in everything which relates to your labours and the work severally, that, independent of other & more unselfish motives any illness of his must cause you much uneasiness. His state of health latterly has not been one to be trifled with and our English writers evidently try him very much.' Prince struggled valiantly on, despite prolonged periods of confinement, administering to the accounts and taking letters, as he had done for over 40 years. He was fighting an uphill battle however, and Gould's son Franklin expressed his concern in a letter to his father in September 1871 that 'Mr Prince still continues very far from well'. The fate of Gould and Prince and that of the company seemed inexorably linked. Gould did all he could to keep them going. 'I remember', said Bowdler Sharpe, 'when Mr Prince was ill, how Mr Gould, though himself an invalid, used to drive out himself every day to see his old friend and to carry with him everything that he fancied would do good to the sufferer.'

Gould by now had been abandoned by all but two of his 'dear Children'. Charles had left London in 1859 at the age of 25 after working for two years at the Institute of Geological Sciences in Exhibition Road to become surveyor general in Tasmania. Although Charles was a consistent correspondent with his family, he had left the Old Country for good, and his father was to see him only once more before he died.

Perhaps a greater sorrow to the ageing Gould was the resignation of his eldest daughter Lizzie from her role as housekeeper. Finally, in 1869, at the age of 33, she married – the only one of Gould's children to do so – a bewhiskered old man almost her father's age. Her father was amazed. She was only, according to the disgruntled Gould, 'exchanging one old man for another'. It was left to her two younger sisters to look after the increasingly difficult and demanding ornithologist.

It was in 1873 that Gould experienced the most devastating bereavement yet to afflict his life. Franklin, his youngest and favourite son, died quite suddenly of a fever on board ship in the Red Sea. By this time a highly qualified doctor, Franklin had accepted the post of medical attendant and travelling companion to the young Lord Grosvenor, the invalid son of the Duke of Westminster. His work had taken him a great deal to Scotland and to Europe, and he had corresponded frequently with his father, describing not only his adventures in the company of his distinguished friend, but noting any ornithological observations en route that might prove of interest to his reader.

The illness that caused Franklin's death came upon him suddenly as he and his charge were returning to England from a trip to India. From the moment it struck, it seems that Franklin resigned himself to death. A graphic and disturbing account given by an eyewitness on board the *Behar* forecast the impact it would have on Franklin's father: 'He began to talk about his brother who had died from a similar fever – the death of that brother had aged his father twenty years and his mind felt distressed to think of the sorrow his own death would cause his Father.'

It was a heavy blow for Gould. Enfeebled by ill health and the pressure of enduring two previous bereavements, Gould was ill-equipped to sustain the shock of a third tragedy. His workshop was facing collapse, and for the first time there seemed no where for him to turn. On 24 March 1873 Sarah wrote a pathetic letter to her eldest sister from Charlotte Street:

My dearest Lizzie,

I have *very* sad accounts of our poor dear Frank to tell you. I hardly know how to tell you the sad news. We had a telegram this evening telling us of the death of our poor dear, dear Frank. He died on board between Aden & Suez – Our poor boy, it is so sad. Poor Father it is a terrible blow to him. *Father*, he seems a sort of broken with it, & says so little, but began to talk this evening of sending for Charlie in case anything happened to him. We cannot seem to realize it at all ourselves, & it will come to one bit by bit. You must not dear Lizzie grieve too much, & we ought all to try & comfort poor Father. No more tonight.

> With our united best love
> > Believe me
> > > Your affectionate sister
> > > > Sai

Charles came back to England for a while at his father's request but left again after a short time on his travels, which eventually took him to South America, never to return. Charles was not prepared to continue in, or anywhere near, his father's business. The competition between father and son might have proved too great for Charles. Gould always shrank from expressing his feelings, and found it even harder to open his heart in times of grief. Their relationship was prone to misunderstandings and unspoken explanations. A friend of Gould's once remarked of the ornithologist, 'He had a really tender and affectionate heart hidden though it was beneath a highly sensitive reserve, which never permitted him the relief of expression. The deaths of his loving wife and two promising boys affected him in a way hardly known outside of his family circle.' Even within the family Gould's emotions remained an enigma, and his inhibitions had a lasting effect upon his children. Charles once remarked in a letter written to his three sisters, who were still living together a few years after their father's death, 'We, as a rule, are not I think a very demonstrative family but probably not a bit less affectionate really than others who profer more.'

Gould's almost complete withdrawal into himself after the death of Franklin greatly affected the formerly bustling and sociable household at Charlotte Street. He tolerated only the company of his two daughters and the occasional visitor. 'He hated the sight of a stranger,' it was said, 'and except the few naturalists of his acquaintance, no one was ever allowed to be admitted to his presence.'

Prince was by now almost bedridden. His wife Amelia was forced temporarily to take over the job of scribe, and wrote her first letter to George Lawrence in New York on 3 September 1873. 'Mr Gould being very unwell', she explained, 'and Mr Prince absent through illness'. Prince occasionally rallied over the next year, as the odd letter dictated by Gould in Prince's hand testifies, but finally died sometime in 1875.

It was the end of an era. Gould's scientific friends were also falling by the wayside and never was Gould more acutely aware of his own mortality. In March 1875 he had two photographic portraits taken of him, a portly, balding, old man with a white beard, by Messrs Maull & Fox, with which to impress posterity.

There was spirit in the old dog yet, however. With Prince gone and his daughter established in his place, and no sign that his illness was to be fatal – or at least not imminently so – Gould rallied, and with all his remaining strength threw himself back into publishing. He had to hand a talented young workmate willing, and pre-eminently qualified, to undertake the demanding job of becoming Gould's assistant. It is possible that until Franklin's death the old ornithologist had held out some hope that his son might continue his work. Franklin – as his father had acknowledged in a touching tribute in the preface to *Birds of Great Britain* (more profoundly expressed one feels than the brief memento to his wife in the preface to *Birds of Australia*) – had worked a good deal with him on the text of his last great opus. The style of the letterpress of *Birds of Great Britain* is markedly more imaginative and romantic in tone – including as reference for example, quotations from Shakespeare, Chaucer, Spencer, and Marlowe – than any of Gould's earlier works. With Franklin's death and Charles' abdication, however, Gould was ready to surrender the privileged position of apprentice to his friend Bowdler Sharpe.

Bowdler Sharpe had first met Gould when Sharpe was a young boy in 1862 'careering about the neighbourhood of Cookham in search of birds'. The little Berkshire village on the Thames was one of Gould's favourite fishing-grounds, and it was here that Sharpe was stopped one day by the ornithologist who questioned him about some wagtails he was carrying. The little boy told him he was taking them to be skinned by the head gardener of a nearby country house, and went on his way. By the time he came back to the gardener's cottage to collect his skins however, Gould had already discovered their where-abouts and requested them for his own collection. The young lad was only too pleased to comply, and Gould, recognising Sharpe showed promise as a bird collector, invited him to visit him in London. Bowdler Sharpe was to become curator of birds at the British Museum, and an 'almost daily visitor' to Gould's house. When he expressed a desire to publish a monograph on kingfishers in the late 1860s Gould lent him 'many specimens' to figure in it.

When he came across the young Sharpe collecting along the banks of the river Thames, Gould possibly felt something reminiscent of his own youth. Whatever his reasons, Gould adopted Sharpe with an openness and familiarity he might have found difficult to show his sons. Sharpe's biographical memoir of his late master, written as an introduction to an analytical index of his works, is generous to a fault. It is in effect an apology for Gould's reputation, and so enthusiastic a eulogy that it comes across more as a whitewash than a genuine portrait of the main. It is clear, however, that the two harboured deep respect for and understanding of one another, and Sharpe a greater patience with the ways of the eccentric old man than many of his older friends and colleagues.

Together Gould and Sharpe set about creating, with Richter and Hart as the principal artists, Gould's stunning collection of the *Birds of New Guinea*, supplemented by contributions from Becceri, d'Albertis and Meyer, and other German and Dutch collectors scouring the islands of the Pacific. Part One, with 13 plates, was published on 1 December 1875.

The announcement of another publication from the legendary ornithologist was greeted with wholehearted applause from the scientific world. 'Every library in Europe has readily given their support to any publication on the newly explored country of New Guinea', Gould told George Lawrence on 12 April 1876, 'the lovely "Birds of Paradise" among the numerous other beautiful things make it one of the most popular of my publications, as always must be the case.' But, he added, 'I am far from well, and write this in pain . . .' To Sir Frederick McCoy he wrote with all his old enthusiasm, 'This New Guinea is a wonderful country teeming with novelties', and once again bringing his persistent drive for subscribers into play, he remarked, 'novelties which bear so much upon Australia as to render it of great interest to you . . .' The work was advertised as much as a continuation of *Birds of Australia* as it was a new publication, thereby casting a net for both the old as well as the new subscriber.

Despite the attraction of the exotic birds it featured, the *Birds of New Guinea* was not so well received in artistic circles as was the *Birds of Great Britain*. The plates were, for some, too fanciful, extravagant, and overworked. Ruskin wrote to his friend the Rev. J.P. Faunthorpe, offering the production for the amusement of his pupils:

'My dear Principal –
    I send you a box to-day containing parts 1-10 and part 12 of Gould's *Birds of New Guinea*. They may serve to astonish some of your little birds, and are only in my way

here. I took them to please the old man, and shall continue
to take them for his sake, sending you the numbers as they
are issued.'

While Bowdler Sharpe was welcomed into the exclusive
world of Gould & Co and inducted into the secrets of the
publishing enterprise, others found it impossible to penetrate
the ornithologist's inner sanctum. The traveller and ornithologist, W.H. Hudson, then in his early 30s, called upon Gould
to see his humming-birds sometime about 1875. His meeting
with the 'pretentious and unscientific ornithologist', or the
'necrologist' as he later liked to call him, was etched with
painful clarity into his memory for the rest of his life. The
sensitive and edgy young man was greeted with patronising
disdain and an affected display of physical agony. 'Gould had
some internal trouble,' says Roberts, 'which afflicted him with
pains he cared not to hide.' His reception left Hudson so bitter
and resentful towards Gould that he vented his anger in a
vicious satirical sketch that appeared in print sometime
afterwards.

Like a character out of Dickens, Gould retreated into a
private world of neurosis and self-obsession, suspicious of
outsiders and fiercely protective of the remains of his not
insubstantial treasures. Mr Reginald Cholmondeley, an
'enthusiastic collector' according to John Guille, the son of the
famous painter Millais, 'paid several visits to Gould at his
house in Charlotte Square [sic]', but it was 'by the exercise of
great tact and patience' that he ultimately 'attained his end,
securing at big prices such specimens as he wanted.' The young
Millais went with Cholmondeley on several of these occasions
and, he says, was 'greatly amused at the old man's veneration
for his treasures, and the tenacity with which he clung to them
when my companion even so much as hinted at a purchase. He
was at this time a confirmed invalid and confined to his couch,
and when a drawer-full of birds was placed in his lap he would

slowly and solemnly lift the lid and handle his specimens with fingers trembling with emotion.'

The old man was determined to plaster over any cracks creeping across the veneer of his reputation before he died; to re-confirm his image not only as the greatest ornithologist that had ever lived, but as the greatest ornithological artist as well. Millais described a pathetic attempt by Gould to impress Millais' father, Sir John Everett Millais, the renowned painter, with his artistic skills. It was a bizarre performance, almost childish in its naiveté and the young Millais watched with fascination as the ageing ornithologist tried to keep up the pretence:

'It was in the middle of winter when my father and I called upon him, by appointment; and after waiting impatiently half an hour in a cold hall, we were just on the point of leaving when the door opened, and we were ushered into his sitting-room. The old man was evidently got up for the occasion. In front of him, as he sat propped up on his couch, was a lovely water-colour drawing of a humming-bird recently discovered (the Chimborazo Hill Star, I think), on which he apparently wished us to believe he was working. But it would not do. We nearly laughed outright when, in reply to an inquiry whether the work was finished, he said, "Oh no; I am just going to put another humming-bird in the background", and suiting the action to the word, sketched on it an object such as never yet was seen on land or sea . . .'

If Gould's artistic fantasy failed to convince the Millais, his birds spoke for themselves: 'And now, calling in his two daughters to help him – for they alone were ever allowed to touch his cases – the old man showed us all his latest gems from New Guinea and the Papuan Islands, and afterwards his unique collection of humming-birds, all of which were set up in cases, and may now be seen (alas! with diminished lustre) in the Natural History Museum at South Kensington.'

The impression the two painters came away with was one that most fairly acknowledged Gould's achievement, despite his failings. 'Artist or not,' they agreed, 'he was a devoted and well-informed naturalist, who by sheer hard work had won his way to the front in a profession in which none but an enthusiast could ever hope to succeed.' The picture of the ageing ornithologist, confined to his couch but transported by his birds, was one that moved the elder Millais to embark upon what many considered his greatest painting. Once again Gould was to be the inspiration of a work of art. 'My father was delighted with all he saw,' said John Millais, 'and on our way home he said to me, "That's a fine subject; a very fine subject. I shall paint it when I have time." And he did.'

'The Ruling Passion' was begun in 1885, four years after Gould's death, and appeared in the Royal Academy Exhibition of that year to rapturous acclaim. 'There are only three things', Ruskin said of the picture shows of 1885, 'worth looking at in them. One is Millais's big picture at the Academy – with the entirely noble old man and the noble young girl in front.' Millais, according to his son, considered it 'one of his best works'. It is indeed a masterpiece of Victorian sentiment – an amiable, bespectacled old man lovingly inspects one of his birds, while a troop of attractive children crowd around him looking on attentively. Scattered about the room with picturesque abandon is a collection of photogenic birds. 'I have never seen any work of modern art with more delight and admiration than this', wrote John Ruskin in 1886.

The painting, however, could not have been further from reality. The engraver, T.O. Barlow, who sat for the principal figure looks nothing like the bearded and balding John Gould; the room, furnished with a big Sheraton bookcase belonging to Mrs Millais, glows warm and welcoming, its central group of figures bathed in divine light. The loneliness, anxiety, and pain that Gould experienced as an old man were anathema to Millais' fanciful imagination. The family with which Millais supplied 'The Ornithologist', as the painting was later called, bears no relation to Gould's own: two professional young models and a little boy in a sailor-suit were substituted for Gould's middle-aged daughters. The cherubic little boys resting next to the old man, peering transfixed and open-mouthed at the revelation in his hands, were Millais' own grandsons, who the artist found irresistible and tended to include – often to the detriment of the picture's composition – in every painting he could. Even the birds were not Gould's own, belonging to Millais' collection.

But Gould would have been pleased, no doubt, to see himself immortalised in this way. To a certain extent it promulgated the myth he had himself managed so convincingly to create. Millais might, now that Gould's guard was down, have seen the strings that governed the great master's movements, but as an artist he was as keen as Gould to idealise the picture and to preserve the illusions of the performance.

'The Ruling Passion' confirmed John Ruskin's romantic sentiments about the ornithologist. Towards the end of his life Ruskin exclaimed to a friend:

'I have made a great mistake. I have wasted my life with mineralogy, which has led to nothing. Had I devoted myself to birds, their life and plumage, I might have produced something myself worth doing. If I could only have seen a humming-bird fly, it would have been an epoch in my life! Just think what a happy life Mr. Gould's must have been – what a happy life! Think what he saw and what he painted . . . You remember that perfectly beautiful picture of Millais' – 'The Ornithologist' – the old man with his birds around him? – one of the most pathetic pictures of modern times.'

For the most part, Gould's *was* a happy life. He adored his work, and it brought him what he longed for most: fame and respectability. His complete surrender to the service of his vocation cushioned him from the blows felt by most people during the course of their lives: it comforted him in times of bereavement; it even assuaged the pain of his illness in his final years. How happy Gould made other people is another question. His children and wife understood and forgave his obsession; colleagues sided passionately either for or against him; artists such as Lear, Audubon, and Wolf felt bitter towards him; and some, those he had crossed in the course of their careers, positively hated him.

Gould's workshop though, was unerringly loyal. His taxidermists, draughtsmen, lithographic printers, colourists, and printers worked for him for years, 'the two latter', as Bowdler Sharpe remarked, 'seem never to have been changed throughout his life.' The secret ingredient that bonded the famous firm of Gould & Co was not so much the charisma of its director as the reliability of his pocket. 'I once asked him what had been the secret of his success in business,' said Bowdler Sharpe, 'and he told me that he had always employed good workmen and had always paid them well. "You should never", said he, "spend fifteen shillings until you have got a sovereign – then you will have five shillings to put by." '

Gould was rarely generous to his colleagues unless it was expedient to be so; but he dutifully provided for the members of his staff and for his employees, assuming a paternalistic role if they were sick and taking pains to safeguard their families. In his will he gave £100 to the widow and daughters of 'my late confidential clerk and assistant Edwin Charles Prince'; and £100 'as a kind of remembrance for the purchase of a ring or any other article that he may prefer in memoriam' to 'my Artist H[blank] C[blank] Richter', whose Christian names Gould still, after 40 years, did not know; and 'to my artist [blank] Hart', (of whose initials, even, after thirty years, he was not aware) 'who has a large family I give a legacy of two hundred pounds'. Gould's son Charles, by comparison, was given 'all of my fishing rods and fishing tackle of every description'. The house in Charlotte Street and all its contents was divided between Sarah and Louise.

Gould's will was signed in April 1878, by which time his health was rapidly deteriorating, but his determination to continue working remained as steadfast as ever. 'I am sorry to say I am still very unwell', he told George Lawrence in January in a letter requesting to see two new humming-birds, 'although able to devote myself to the pursuit of my favourite study.' To J. Noble, a prospective subscriber, he wrote, on 17 October that he was 'still on the sofa, generally bearing my pains with the fortitude I trust of a Christian, I may or may not, get better, the doctors say I shall.' But, he added, 'The solitude in which I live, is greatly solaced by the occupation I find in my publications, particularly in those wonderful things from New Guinea, a country differing from all others in its ornithological productions.' A few days later he wrote again to Mr Noble, who had been successfully enlisted as a client, assuring him that, 'By hook or crook you shall have a complete set of my works. If you leave this to me I will set about the task which will be but a light one. Will you excuse a pencil note which I write on my back. If I attempt this with ink the fluid obeys the laws of nature by leaking downward.' To Bowdler Sharpe, Gould revealed a little more concern. On 7 June he wrote, 'I send your cheque that I may keep faith with you. In one of the papers you speak of a new *Polyplectron* having been discovered in Borneo as you often have [?] to me, with a promise that I shall see *it*, but alas – I have never yet set eyes on so interesting a bird as a new polypl. Shall I ever see it. J. G.'

In the same year the first meeting of the Committee of the British Ornithologists Union was held to revise the nomenclature of British birds. Despite, or maybe because of, being the most famous ornithologist in England at the time,

Gould had not been invited to join the British Ornithologists Union at its inception in the 1850s, or even since. It had clearly been a deliberate attempt on the part of Alfred Newton and the other Cambridge-based ornithologists to cut Gould down to size, and, although there is no mention of it in any existing correspondence, this must have hurt him considerably. Gould frequently contributed papers to the Union's quarterly journal the *Ibis*, which in turn reviewed his works with unqualified enthusiasm; but he was not to be accepted by the University as one of the mainstream ornithological scientists of the day. When it came to reassessing the designation of British bird names, Gould's classifications were singled out for particular criticism, a fact that might explain his earlier exclusion from the Union. A new breed of ornithologists was coming to the fore, who were more professional in their approach, and who were sceptical about the methods and discoveries of what had become a 'gentleman's hobby'. They were keen to disassociate themselves from anything that was not purely scientific in nature; John Gould by dint of his glittering rank as an illustrator and publisher was consequently considered less significant as a scientist. He was a victim of his own success: he could not be allowed to have it both ways.

Gould was one of the last of the old school of ornithological illustrators, dextrously combining the disciplines of science and art. As the study of science specialised further and the new branch of biology began to break forth, an irrevocable wedge was driven between the two, and never again would publications such as Gould's be tolerated with such equanimity. The day of the luxury imperial folio was almost over. Gould's works would not measure up to the new standards of practicality set by science, and as works of art they could not stand alone. Gould's favourite species, and his most famous and extravagant illustrative subject – his humming-birds – was coming under special attack, and for the first time he began to fear being ousted by the young cuckoos appearing in his nest.

D.G. Elliot, who was then working on his own humming-bird synopsis, explained the situation to R. Ridgway in March 1878:

'When in London I sent a description of a new species of Ionolaema to the Ibis which will appear in the next number. It is by far the most splendid member of the genus, and raises the number of this fine group to four. It was in Gould's collection but he did not dare describe it. In fact since I began my critical articles in the Ibis he has been very shy of humming-birds, as he finds his Monographs is exceedingly vulnerable. He thought more of publishing plates for the profit they brought him than of the value of the species he described, and he finds these "bogus" creations are coming to plague him disagreeably, so as in the case of Ionolaema, he fears to describe a really distinct form when he has it. I regret to say he is very ill, and utterly unfit for any kind of work, even that of picture making.'

If this kind of attack from the 'young upstarts' of his profession was bad enough, how much more painful would Gould have felt their pity, as evinced in another letter from Elliot to Ridgway a year later:

'When you get my Synopsis you will see I have rejected "Cometes", because it was previously employed by Hodgson some six years before Gould used it. The proper term for the "fire-tails" is Sappho. "Gouldia" I have employed in spite of its having been preoccupied for the reason that I think it should have a rather different consideration given it than is accorded to an ordinary generic term. It is designed as a compliment to a leading naturalist, and I think ought to be allowed to stand, even, (as is mostly likely in the present case) if its proposer did not know that it had been already employed. If it should be

decided that it must give way, I do not care to be the one
to initiate its downfall.'

Gould could not take this kind of presumption lying down, and
despite his infirmity and declining strength, launched into
counter-attack. On 1 August 1880, 'undeterred by serious
illness', he published the first part of *A Monograph of the
Trochilidae, or Family of Humming-birds, Supplement*, and
scheduled another publication, *A Monograph of the Pittidae or
Ant-Thrushes of the Old World*, for 1 October 1880. Hotly,
Gould denied his illness had incapacitated him, however
absurd his assertions may have seemed to those who knew the
truth. On 9 December 1879, in a letter written by Amelia
Prince to a correspondent in Amsterdam, he claimed, 'I have
been steadily working, whilst lying ill on my back, and have
figured more new things than in a quarter of my life before.'

In fact Gould was by now hardly capable of writing, and on
8 September 1880 he scrawled a short note of instruction – the
last known to have been written by him – to Bowdler Sharpe
saying that he was 'In great pain. My opiate leaves me . . . I
write this and all other notes on birds on my bed, so excuse
them.' He died five months later on 3 February 1881 at the age
of 76. Bowdler Sharpe was invited the following day by Sarah
Gould 'as an old friend of my Father's & as one who really liked
& respected him' to attend his funeral along with the rest of the
family. Sharpe took the place of Charles, who stayed away in
the Far East.

When it came to the problem of the publications still
unfinished – the *Birds of Asia* had between two to three parts to
go; *Birds of New Guinea* was only 11 parts into its total of 25
when Gould died; another four parts of the humming-bird
supplement needed to be published; and a total of four parts
had originally been expected for the *Monograph of Pittidae*,
although only the second part appeared after Gould's death,
Bowdler Sharpe was the obvious and natural person to don

Gould's mantle. 'I congratulate you' wrote Tomasso Salvadori from Turin to Gould's erstwhile protégé, 'with having been charged of [sic] the continuation of Gould's works, and which [wish] that these works will give you some of the material profit which Gould had.'

Gould had indeed made a fortune in his lifetime. No other illustrator or ornithological publisher came close to the success Gould made of his business. The single copy, in parts, of his successive works, which he had been bound by the copyright act to present to the British Museum, represented a value of nearly £700. Over a year before his death Gould had the satisfaction of recognising that, 'the publications are more and more liked and many of the very large ones are entirely sold and at a premium.' Sotheran's, the Booksellers of Piccadilly, who purchased the remaining stock of Gould's work, announced barely three years after the author's death that 'of several [of his series] just one copy remains for sale while of the others on hand only a few copies of each remain.'

More than the 'material profit' the ornithologist enjoyed during his lifetime, Gould's publications confirmed a greater objective – that of immortalising the name of Gould. Alfred Newton wrote of 'the marvellous series of works by which the name of John Gould is likely to be always remembered . . . the wonderful tale consisting of more than forty folio volumes and containing more than 3000 coloured plates'. *The Zoologist* acknowledged that, 'In this large series of beautiful volumes Mr. Gould has certainly raised an enduring monument to his own fame'. The Duke of Westminster assured Gould's daughters, 'he has left a permanent pleasure, in his delightful books, for thousands and an enduring monument of his own industry and ability.'

Gould's personal library copy of the complete works, bound mostly in green morocco, encased in a special glass-fronted bookcase seven feet long, his 'portrait recently painted in oils', and all his 'Ornithological and other oil paintings absolutely' were bequeathed to his eldest daughter, Lizzie. 'My wish', he declared, 'is that they shall all be kept in the family in the nature of heirlooms'. They represented the official testament of his success and his legacy to posterity. Sadly, Gould's own set of his works is no longer in the family. It was sold in 1987 for £360,000 and the volumes scattered all over the world. Some went to private collectors in Europe, America and Saudia Arabia; but some, like his *Birds of Australia* set, were bought by dealers who cut them up for the plates.

Gould's remaining bird collection, containing his vast selection of 5,378 humming-birds and 7,017 other skins of different groups of birds, such as his famous trogons, toucans, birds of paradise, and pittas, 'all unrivalled for completeness and condition' was bought by the British Museum for £3,000. The 'disgraceful spectacle' of Gould's 'unrivalled' Australian collection sailing off to the States was not to be repeated. Many of his eggs, too, were retrieved by the British Museum to form a collection of over 1,700.

Gould had distinguished himself during his long, hard and ambitious career as a pioneer of the new science of ornithology. He was one of the foremost collectors of the world; a first-rate taxidermist and taxonomist; one of the finest publishers in London; and a mean, if sometimes unscrupulous, business-man. The ornithologist's most fundamental wish, however – a wish that was egotistical yet also engagingly frank – was to be remembered simply for his love of birds. The gardener's son, in his own words, wished his epitaph to read: 'Here lies John Gould, the "Bird Man".'

Sources

Page 10   *'Botany is not my forte.'*, J. Gould to Sir W. Jardine, 19 Jan. 1841; Roy. Mus. Scot., Edinburgh

Page 12   *'Gould's artistic talent . . .'*, obituary by P. L. Slater, 'Proc. of the Roy. Soc. London', 33 (1882), pp.17-19

CHAPTER 2: **Birds of a Feather**

Page 14   *'Rome, at the period. . . .'*, Original prospectus of the proposed Zoo. Soc., 1 March 1825; J. Bastin, 'The first prospectus of the Zoo. Soc. of London: new light on the Society's origins', *J. Soc. Biblphy. Nat. Hist.*, 5 (5) (1970), 369-88

Page 15   *'Contemplates a more practical cult. . . .'*, Mr Bicheno, Zoological Club President, annual address to Zoological Club of the Linnean Soc., 29 Nov. 1826, [*see* Scherren p.24]

Page 16   *'In the first instance . . .'*, Stamford Raffles to Sir R.H. Inglis, 28 April 1825, [*see* Scherren p.17]

Page 17   *'My friend Mr Gerrard . . .'*, P.B. Sharpe, *An analytical index to the works of the late John Gould, FRS*, pp.x-xi, (London, Henry Sotheran & Co., 1893)

'An Address delivered at the sixth and last Anniversary Meeting of the Zoo. Club of the Linnean Soc. of London on the 29th of November, 1829', N.A. Vigors, Esq. AM FRS, *et al.*, *Mag. of Nat. History*, May 1830, pp.201-26

Page 18   *'Gould is a man of great industry. . . .'*, J.J. Audubon to Rev. J. Bachman, 20 April 1835, H. Corning, *Letters of John James Audubon 1826-1840*, (2 vols., Boston, 1930), pp.67-70

Page 21   *'By the way Gould always sends me. . . .'*, A.F.R. Wollaston, *Life of Alfred Newton*, (London, John Murray, 1921)

Page 22   Obituary of J. Gould by P.L. Sclater, Proceedings of the Roy. Soc. of London, 33 (1882), pp.7-19

CHAPTER 3: **The Start of a Name**

Page 24   Palmer, *Life of Joseph Wolf*, (1895), p.71

E. Coxen, 18 James Street, to her mother, undated, (*c*.

autumn 1827); G. Edelsten

Page 26   J. Gould to Sir W. Jardine, 1 Sept. 1830; Roy. Mus. of Scot., Edinburgh

Page 27   Obituary of J. Gould, *The Times*, 6 Feb. 1881

'*Gould is about to publish. . . .*', P.J. Selby, Twizell House, to Sir W. Jardine, 30 Nov. 1830; Balfour & Newton Lib. Zoo. Dept., Cambridge Univ.

Page 28   '*I have this day forwarded . . .*', J. Gould to Sir W. Jardine, 20 Dec. 1830; Roy. Mus. of Scot., Edinburgh

Page 29   '*Have you got Gould's nos 4 & 5. . . .*', P.J. Selby to Sir W. Jardine, 26 April 1831, Balfour & Newton Lib., Zoo. Dept., Cambridge Univ.

R. Havell to J.J. Audubon, 30 Nov. 1831; A. Ford, *John James Audubon*, (Norman, OK, Univ. of Okla. Press, 1964), p.292

'*I have forwarded. . . .*', J. Gould to N.A. Vigors, 5 Nov. 1833; Roy. Mus. of Scot., Edinburgh

Page 30   '*By a parcel from London . . .*', Sir W. Jardine to P.J. Selby, 15 Nov. 1831; Balfour & Newton Lib., Dept. of Zoo., Cambridge Univ.

Page 31   J. Gould to B.H. Hodgson, 6 March 1837; Zoo. Lib., Nat. Hist. Mus., London

'Remarks & References to Gould's Letter of 6th March 1837.' Undated. B. Hodgson to his son, B.H. Hodgson; Zoo. Lib., Nat. Hist. Mus., London

Page 32   '*Surely it is too much. . . .*', B. Hodgson, Canterbury, to J. Gould, 10 March 1837; Zoo. Lib., Nat. Hist. Mus., London

CHAPTER 4: Lear's Years of Misery

Page 34   '*He was one I never really liked . . .*', E. Lear, 7 Feb. 1881, V. Noakes, *Edward Lear – The Life of a Wanderer* (Collins, 1968), pp.40-1

Page 35   '*colouring prints . . .*', E. Lear to C. Fortescue, 1 Jan. 1863, Strachey, *Letters of Edward Lear* (*c.* 1907), p.xxviii

Page 38   '*Should you come to town. . . .*', E. Lear to C Empson, 1

Oct. 1831, V. Noakes (ed.), *Edward Lear – Selected Letters* (Oxford Univ. Press, 1988)

Page 39   *'Sir I received yesterday. . . .'*, W. Swainson, St Alban's, to E. Lear, 2 Nov. 1830, B. Reade, *Edward Lear's Parrots* (Duckworth, 1949)

Page 40   Sclater, *op. cit.*, 33: pp.17-19

'the first book of the kind. . . .', Lear to the Earl of Northbrook, 11 Oct. 1867, V. Noakes (ed.), *op. cit.*

'My reasons. . . .', E. Lear to C. Empson, 1 Oct. 1831, V. Noakes (ed.), *op. cit.*

Page 41   *'Respecting my Parrots. . . .'*, E. Lear to Sir W. Jardine, 23 Jan. 1834; Roy. Mus. of Scot. Edinburgh

Page 42   *'You will perceive. . . .'*, J. Gould to Sir W. Jardine, 23 Jan. 1834; Roy. Mus. of Scot., Edinburgh

J.J. Audubon to Rev. Bachman, 20 April 1835, H. Corning, *Letters to John James Audubon 1826-40*, 1930, p.67

Page 44   *'Gould has been so clamorous. . . .'*, E. Lear to Sir W. Jardine, 11 March 1836; Roy. Mus. of Scot., Edinburgh

'I got your letter . . .', E. Lear to J. Gould, 31 Oct. 1836; Houghton Lib., Harvard

Page 45   *'this lampblack & grease work'; 'all my daylight hours'*, E. Lear to J. Gould, 14 Oct. 1863; Houghton Lib. Harvard

'Lord! Lord! How cross you used to be . . .', E. Lear to J. Gould, 15 March 1838; Ellis Collection, Spencer Lib. Univ. of Kansas, MS P498:1

Page 46   J.J. Audubon to Rev. Bachman, 20 April 1835, H. Corning, *op. cit*, pp.67-70

'far far too complimentary. . . .', E. Prince, Broad Street, to J. Gould, 2 March 1839; Mitchell Lib., Sydney, MS A173

Page 47   E. Lear, 1863, V. Noakes (ed.), *Selected Letters of Edward Lear*, Biographical Register, p.xxviii

CHAPTER 5: **A Dram-Bottle from Darwin**

Page 49   *'Would not a work on the Birds of Australia. . . .'*, J. Gould to Sir W. Jardine, 2 Nov. 1836; Balfour &

Newton Library, Cambridge Univ.

Page 51   *'I must not omit to tell you . . .'*, J. Gould to Sir W. Jardine, 16 Jan. 1837; Balfour & Newton Library, Cambridge, Univ.

Page 56   C. Darwin, *Ornithological Notes*; Univ. Lib. Cambridge, Handlist of Darwin Papers No. 29(ii) Birds, Galapagos DAR 29.3f73, N. Barlow (ed.), 'Darwin's Ornithological Notes', *Bull. of the Brit. Mus. (Nat. Hist.) Hist. Series*, vol.2, no.7, p.262

Page 57   Darwin's MS record of John Gould's species designations; Cambridge Univ. Lib., DAR 29:3 (f.27)

Page 59   *'It is most amusing to see. . . .'*, A. Newton, Cambridge, to E. Newton, 25 April 1864, A.F.R. Wollaston, *Life of Alfred Newton* (London, John Murray, 1921)

Page 60   C. Darwin to E. Prince, 21 Jan. 1839; NY Botanical Garden Lib.

            'I received with much pleasure . . . J. Gould to C. Darwin, 13 April, 1838; Zoo. Lib., Nat. Hist. Mus., London

CHAPTER 6: **Birds of Passage**

Page 62   J.J. Audubon to Rev. Bachman, 14 Aug. 1837, H. Corning *op. cit.*, vol.2, p.176

            'You are perhaps aware . . .', J. Gould to W. Swainson, 21 Jan. 1837; Linnean Soc., London

Page 63   *'Mr Yarrell has just called . . .'*, E. Prince to J. Gould, 2 May 1839; Mitchell Lib., Sydney

            'The figures of the heads . . .', P.J. Selby to Sir W. Jardine, *c.* Feb./March 1837; Balfour & Newton Lib., Cambridge Univ.

            'Gould is publishing the Birds of Australia . . .', J.J. Audubon to Rev. Bachman, Oct 4, 7 1837, H. Corning, *op. cit.*, vol.2, pp.185-6

Page 64   *'You will perceive. . . .'*, J. Gould to Sir W. Jardine, 2 July 1841; Roy. Mus. of Scot., Edinburgh

            'the cancelled parts. . . .', J. Gould to Lord Derby, 5 Feb. 1844, Wagstaffe & Rutherford, *Letters from Knowsley Hall, Lancashire*, no.17

Page 65   *'I wish other men of science . . .'*, Gov. J. Hutt to J. Gould, 23 April 1839; Roy. Soc. Tasmania Backhouse Collection MS RS58.

'Mr Gould the author of Birds of Europe . . .', J.J. Audubon to Rev. Bachman, 14 April 1838, H. Corning, *op. cit.*, vol.2, p.202

Page 66   *'I and Joseph are much pleased'*, E. Prince to J. Gould, 5 Feb. 1840; Mitchell Lib., Sydney

Page 67   *'sudden & severe indisposition of Mrs Gould'*, E. Prince to Sir W. Jardine, 10 Aug. 1838; Roy. Mus. of Scot., Edinburgh

J. Gould's power of attorney. Signed 7 May 1838; Mitchell Lib., Sydney

Page 68   *'Prince has been with me from the commencement . . .'*, Preface to *Birds of Australia*, June 1848

Page 69   *'I am so busy . . .'*, J. Gould to Sir W. Jardine, 30 April 1838; Balfour & Newton Lib., Cambridge Univ.

Page 71   J. Gould to the Chair. of the Sci. Comm. Zoo. Soc. London, 20 June 1838; Mitchell Lib., Sydney

Page 72   *'He writes in high spirits . . .'*, E. Prince to Sir W. Jardine, 10 Aug. 1838; Roy. Mus. of Scot., Edinburgh

'we are delighted . . .', E. Prince to J. Gould, 18 Aug. 1838; Mitchell Lib., Sydney

J. Gould to the Chair. Sci. Comm. the Zoo. Soc. London, 10 May 1839, Pub. in Proc. of the Zoo. Soc., 7:*139-45*

Page 73   Soft-plumaged Petrel, *Birds of Australia*, vol.vii, p.150; *Handbook to Birds of Australia*, vo.ii, sp.631

CHAPTER 7: The Harvest Meets the Sickle

Page 76   E. Gould, Hobart, to her mother, Broad Street, 8 Oct. 1838; Mitchell Lib., Sydney.

Page 77   E. Gould, Hobart, to her mother, 10 Dec. 1838; Mitchell Lib., Sydney

Page 78   Diary of Lady J. Franklin, G. Mackaness, *Some private correspondence of Sir John and Lady Jane Franklin*, (Sydney, privately printed, 1947), Part I, pp.39-53

Page 80 J. Gould, Recherche Bay, to E., Hobart, 20 Dec. 1838; Mitchell Lib., Sydney

Page 82 E. Gould, Hobart, to her mother, 3 Jan. 1839; Mitchell Lib., Sydney

Page 84 J. Gould, George Town, to E. Gould, Hobart, 8 Jan. 1839, Mitchell Lib., Sydney

Page 87 *'I shall always feel . . .'*, J. Gould to Sir W. Jardine, 9 Sept. 1840; Balfour & Newton Lib., Cambridge Univ.

'Tell Lady Franklin. . . .', *'I am now in Capt. Friend's office . . .'*, J. Gould, George Town, to E. Gould, Hobart, 20 Jan. 1839; Mitchell Lib., Sydney

Page 88 E. Prince to Sir W. Jardine, 20 July 1839; Roy. Mus. of Scot., Edinburgh

Page 89 E. Prince to J. Gould, 2 March 1839; Mitchell Lib., Sydney

Page 90 E. Prince to J. Gould, 20 May 1839; Mitchell Lib., Sydney

E. Gould to her mother, 9 Jan. 1838 (error for 1839); Mitchell Lib., Sydney

Page 91 E. Gould to her mother, 15 Feb. 1839; Mitchell Lib., Sydney

J. Gould, Hobart, to Capt. J. Washington, London, 14 Feb. 1839; Roy. Geog. Soc., London

CHAPTER 8: **Adventure in the Newest World**

Page 92 White-headed petrel, *Birds of Australia*, vol.vii, pl.49, *Handbook to Birds of Australia*, vol.ii, sp.630

Page 93 J. Gould to Sir J. Franklin, 25 Feb. 1839; Scott Polar Res. Instit., Cambridge

Page 94 Lady J. Franklin's Diary, 4 July 1839, Sydney; Nat. Lib. of Aust., Canberra

Mrs N.G. Sturt, *Life of Charles Sturt* 1899 (*see* Cleland, 1937, *Emu* 36:205)

Page 95 *'You, Sir, fully understand . . .'*, J. Gould to Sir J. Franklin, 25 Feb. 1839; Scott Polar Res. Instit., Cambridge

Stephen Coxen's suicide, *Sydney Morning Herald*, 5 Sept. 1844

Page 96 J. Gould, River Hunter, to Eliz. Gould, Hobart, 20 March 1839; Mitchell Lib., Sydney

E. Gould to Mrs Mitchell, 28 May 1839; Mitchell Lib. Sydney

Page 97 Lyre-bird, *Birds of Australia*, vol.iii, pl.14

Page 98 Satin bower-bird, *Birds of Australia*, vol.iv, pl.10

J. Gould to Mrs Owen, 'Monday' (undated); Univ. Lib., Cambridge, Add.5354.205

Page 99 J. Gould, River Hunter, to E. Gould, Hobart, 20 March 1839; Mitchell Lib., Sydney

Page 100 E. Gould, Hobart, to Mrs Mitchell, Broad St, 30 May 1839 (postscript to letter, 28 May 1839); Mitchell Lib., Sydney

E. Gould to Mrs Mitchell, 28 May 1839; Mitchell Lib., Sydney

Page 101 J. Gould, Adelaide, to E. Prince, Broad St, 30 June 1839, *Mag. Nat. Hist*. 3 (new series):568

Page 102 Porphyry-crowned lorikeet, *Birds of Australia*, vol.v, pl.53

Page 103 J. Gould, Maitland, to the Zoo. Soc. London, 28 Sept. 1839, *Ann. Nat. Hist*., Jardine, vol.v, 1840 pp.117-19

'*He has sent home. . . .*', E. Prince to Sir W. Jardine, 1 Jan. 1840; Roy. Mus. of Scot., Edinburgh

Page 104 '*I have visited South Australia. . . .*', J. Gould to Sir W. Jardine, 28 Sept. 1839; Balfour & Newton Lib., Cambridge Univ.

Dr Bennett (containing quote from J. Gould's letter), Sydney, to Prof. Owen, 23 June 1839; Roy. Coll. of Surg., London

Harlequin bronzewing, *Birds of Australia*, vol.v, pl.66; *Handbook to Birds of Australia*, vol.ii, sp.464

Page 105 C. Sturt to J. Gould, 19 May 1843, Wagstaffe & Rutherford, *Letters from Knowsley Hall, Lancashire*, (North Western Naturalist, 1954-5)

Page 106 J. Gould to Lord Derby, 22 April 1844, Wagstaffe & Rutherford

Page 107 'an orderly arrived with dispatches . . .', E.S. Mahoney, 'The first settlers at Gawler', Proceedings Roy. Geog. Soc. of Australasia: South Australian Branch, 28 (1928):53-82

CHAPTER 9: **A Glorious Haul**

Page 109 E. Prince to Sir W. Jardine, 3 March 1840; Roy. Mus. of Scot., Edinburgh

'The case on its arrival . . .', E. Prince to J. Gould, 29 June 1839; Mitchell Lib., Sydney

'I yesterday got out the Box . . .', E. Prince to J. Gould, 5 Feb. 1840; Mitchell Lib., Sydney

Page 110 'I regret you should have determined . . .', E. Prince to J. Gould, 5 Feb. 1840; Mitchell Lib., Sydney

J. Gould to Gilbert, 16 Nov. 1840; Nat. Lib. of Aust., Canberra

Page 111 'My thanks are due . . .', E. Prince to Sir W. Jardine, 7 April 1840; Roy. Mus. of Scot., Edinburgh

'John, of course . . .', E. Gould to her mother, 20 Aug. 1839; Mitchell Lib., Sydney

Page 112 E. Prince to J. Gould, 5 Feb. 1840; Mitchell Lib., Sydney

Page 113 E. Gould's diary, 20 Aug.-29 Sept. 1839; Item A1763, Mitchell Lib., Sydney

Page 114 J. Gould, River Hunter, to Sir W. Jardine, 28 Sept. 1839; Balfour & Newton Lib., Cambridge

Page 116 E. Gould's diary, *loc. cit.*

J. Gould to Sir J. Franklin, 2 Nov. 1839; Scott Polar. Res. Instit., Cambridge

E. Gould, Yarrundi, to her mother, 6 Dec. 1839; Mitchell Lib., Sydney

Page 117 Harlequin bronzewing, *Birds of Australia*, vol.v, pl.66; *Handbook to Birds of Australia*, vol.ii, sp.464

Page 118 'the black fellows . . .', J. Gould to E.P. Ramsay, Sydney, 26 Nov. 1866; MS 1217, Nat. Lib. of Aust., Canberra '*extreme cheerfulness of disposition*'; '*naturally made me anxious . . .*', Warbling grass-parakeet, *Birds of Australia*, vol.v, pl.44; *Handbook to Birds of Australia*, vol.II, sp.439

Page 119 Mrs Owen's diary, 27 March 1841, Rev. R. Owen, *The Life of Richard Owen By His Grandson*, 2 vols. (London, John Murray, 1894)

Page 120 Crimson chat, *Handbook to Birds of Australia*, vol.I, sp.233

'*Of its mode of nidification . . .*', Nightjar, *Ibid.* vol.I, sp.40

Square-tailed kite, *Ibid.*, vol.I, sp.22; *Birds of Australia*, vol.I, pl.22

Page 122 '*I am happy to say . . .*', J. Gould, Sydney, to Sir J. Franklin, 6 April 1840; Scott Polar Res. Instit., Cambridge

Page 124 '*My collection . . .*', J. Gould to Sir W. Jardine, 9 Sept. 1840; Newton Lib., Cambridge

J. Gould to C. Bonaparte, 2 Feb. 1841; Musée Nat. d'Hist. Nat., Paris

CHAPTER 10: The Sacrifice to Science

Page 127 J. Gilbert, Sydney, to J. Gould, 4 May 1840; Mitchell Lib., Sydney

Page 128 J. Hutt, Gov. of West. Aust., to J. Gould, 23 April 1839; Roy. Soc. of Tasmania Backhouse Collection MS RS58

Page 129 Mrs R. Brockman's description of J. Gilbert, A.J. Campbell, *Nests and Eggs of Australian Birds* (Introduction), (Sheffield, Pawson & Brailsford, 1900)

Page 130 J. Gilbert, Sydney, to J. Gould, 4 May 1840; Mitchell Lib., Sydney

Page 131 '*The Tent you spoke of . . .*', J. Gilbert, Sydney, to J. Gould, 8 June 1840; Mitchell Lib., Sydney

Page 132 '*You would have been highly amused. . . .*', J. Gilbert, Port Essington, to J. Gould, 18 July 1840; Mitchell Lib., Sydney

'*I have collected . . .*', J. Gilbert, Port Essington, to J.

Gould, 19 Sept. 1840; Mitchell Lib., Sydney

Page 133   *'I received both your letters . . .',* J. Gould to J. Gilbert, Sydney, 5 April 1841; Nat. Lib. of Aust., Canberra

General instructions for J. Gilbert from J. Gould, 21 Jan. 1842; Mitchell Lib., Sydney

Page 134   *'collect skins of all kinds . . .',* J. Gould's notes and instructions for J. Gilbert, *c.* 20 Jan. 1842; Mitchell Lib., Sydney, and Nat. Lib. of Aust., Canberra

Page 135   J. Gilbert, on board the *Houghton le Skerne,* to J. Gould, 5 Feb. 1842; Mitchell Lib., Sydney

Page 136   *'From Johnson Drummond . . . quite new',* J. Gilbert, Perth, to J. Gould, 9 Oct. 1842, Wagstaffe & Rutherford, *Letters from Knowsley Hall, Lancashire*

Page 137   *'You know with what delight . . .',* J. Gilbert, Perth, to J. Gould 13 Sept. 1842, (Wagstaffe & Rutherford, *op. cit.*)

Page 138   J. Gilbert, Darling Downs, to J. Gould, 8 June 1844; (from copy written by 13th Earl of Derby, 27 Nov. 1844), Liverpool City Lib., accession no. 920, DER (13) 1/67/11

'something as will give an eclat . . .'; '. . . accomplished our objects', J. Gilbert, Darling Downs, to Dr G. Bennett, 10 Sept. 1844; Leichhardt Papers C161, Mitchell Lib., Sydney

Page 139   J. Gilbert, Swan River, to J. Gould, 3 Sept. 1839; Mitchell Lib., Sydney

'I am happy to say . . .', J. Gilbert, Port Essington, to J. Gould, 19 Sept. 1840; Mitchell Lib., Sydney

Page 140   *'I freely confess. . . .',* J. Gilbert, Darling Downs, to Dr G. Bennett, Sydney, 10 Sept. 1844; Leichhardt, Papers C161, Mitchell Lib., Sydney

'Your journey. . . .', J. Gould to L. Leichhardt, 12 July 1845; Leichhardt Papers C161, Mitchell Lib., Sydney

J. Gould to Fairfax, 16 Dec. 1845; Mitchell Lib., Sydney

Page 141   *'I have just learnt . . .',* J. Gould to Dr G. Bennett, Sydney, 16 Dec. 1845; Mitchell Lib., Sydney

J. Gould to J. Stokes, West. Aust., 2 Feb. 1846; Mitchell Lib., Sydney

J. Gould to Sir W. Jardine, 1 April 1846; Balfour & Newton Lib., Cambridge

J. Gould to Dr G. Bennett, 1 June 1846; Mitchell Lib., Sydney

J. Roper to J. Gould, 12 May 1846, 'Proc. Zoo. Soc. London', 1846, pp.79-80

Page 144 J. Gilbert's journal of the Leichhardt expedition; Mitchell Lib., Sydney

Page 146 *'How could Gould....'*, L. Leichhardt, Sydney, to Dr G. Bennett, 2 Sept. 1847; Mitchell Lib., Sydney

CHAPTER 11: **A Double Defection**

Page 147 *'I fear I also shall not . . .'*, J. Gould to Sir W. Jardine, 2 July 1841; Roy. Mus. of Scot., Edinburgh

'Mrs G. I assure you. . . .', J. Gould to Sir W. Jardine, 2 July 1841; Roy. Mus. of Scot., Edinburgh

Page 148 *'Poor Mr Gould . . .'*, E. Prince to Sir W. Jardine, 17 Aug. 1841; Roy. Mus. of Scot., Edinburgh

'You will however . . .', J. Gould to E. Lear, (draft), 24 Aug. 1841; G. Edelsten

'You will far more readily . . .', E. Prince to Sir W. Jardine, 17 Aug. 1841; Roy. Mus. of Scot., Edinburgh

'For the sake of my Dear Little Children . . .', J. Gould to E. Lear (draft), 24 Aug. 1841; G. Edelsten

'You will my dear Sir William. . . .', J. Gould to Sir W. Jardine, 1 Oct. 1841; Roy. Mus. of Scot., Edinburgh

Page 149 E. Gould Muskett Moon, 'Some pages of reminiscences of her early life', M. Lambourne

C. Sturt to J. Gould, 19 May 1843, Wagstaffe & Rutherford, *Some Letters from Knowsley Hall, Lancashire*, 1954-1955

'She will be a grievous loss . . .', P.J. Selby to Sir W. Jardine, 30 Aug. 1841;' Roy. Mus. of Scot., Edinburgh

J. Gould to H. Strickland, 13 Oct. 1841; Balfour & Newton Lib., Cambridge Univ.

Page 150   *'the loss I have sustained. . . .'; 'the loss of my very efficient
. . .'*, J. Gould to Sir W. Jardine, 1 Oct. 1841; Roy. Mus.
of Scot., Edinburgh

Page 151   *'I shall soon be more at liberty . . .'*, J. Gould to Sir W.
Jardine, 11 Aug. 1842; Balfour & Newton Lib.,
Cambridge Univ.

'I myself look forward . . .', J. Gould to Lord Derby, 10
Jan. 1841 (error for 1842), Wagstaffe & Rutherford,
Letters from Knowsley Hall

'By the bye I dreamed . . .', J. Gould to J. Gilbert, 14 July
1842 (draft); Nat. Lib. of Aust., Canberra

Page 152   *'Having many years paid . . .'*, J. Gould to Sir W.
Jardine, 2 Dec. 1842; Balfour & Newton Lib.,
Cambridge Univ.

'Mr Gould is on the Continent. . . .', E. Prince to Sir W.
Jardine, 8 Aug. 1843; Roy. Mus. of Scot., Edinburgh

'During my last Continental trip. . . .', J. Gould to Sir W.
Jardine, 10 Oct. 1843; Newton Lib., Cambridge

'With respect . . .', J. Gould to Lord Derby, 5 Feb. 1844,
Wagstaffe & Rutherford, *op. cit.*

Page 153   *'1800 specimens in all . . .'*, B. Sharpe, *op. cit.*, p.xviii

J. Gould to E. Wilson, 8 May 1847, R.M. de
Schauensee, 'On some avian types, principally Gould's
in the collection of the Academy', 109 (1957):123-246;
Proceedings of the Acad. Nat. Sci. of Phila. Lib. of
Acad. of Nat. Sci., Phila.

Page 154   Sir W. Jardine to E. Wilson, 9 July 1847, R.M. de
Schaunsee, *op. cit.* Lib. of Acad. of Nat. Sci., Phila.

Page 155   B. Sharpe, *op. cit.*, pp.xviii-xix

CHAPTER 12: **All That Glitters**

Page 157   *'That our enthusiasm . . .'*, A Monograph of the
Trochilidae, or Family of Humming-birds, (preface), 1861
The Humming Birds Ridgway, 1890), p.254

Page 158   *'this gentleman and myself . . .'*, A Monograph of
Trochilidae, (preface), 1861 (Ridgway, 1890), *op. cit.*,
p.254

Page 159 *'I have now been confined. . . .'*, J. Gould to Sir W. Jardine, 19 Feb. 1846; Balfour & Newton Lib., Cambridge Univ.

'Gould has visited France. . . .', E. Prince to Sir W. Jardine, 30 July 1846; Roy. Mus. of Scot., Edinburgh

Page 160 *'Both Frenchmen and Belgians. . . .'*, *A Monograph of the Trochilidae* (Introduction) (Ridgway), *The Humming Birds*, p.255

Page 161 Letter requesting specimens (*'To Whom it may Concern'*) from J. Gould, *c*. 16 July 1869; Harvard Univ. Lib.

Page 162 W.L. Baily *'Illustrations of the Trochilidae or Humming Birds'*, 1855-8, Phila., Unpub., Lib. of Acad. of Nat. Sci., Phila.

E. Coues, *Ornithological bibliography* (Government Printing Office, 1878-80, Washington D.C.), Part 3, Addendum to *Trochilidae*

Page 163 *'I am extremely much obliged . . .'*, C. Darwin to J. Gould, 6 Oct. 1861, Cambridge Univ. Lib. Add.4251.330

Page 164 *'Humbolt has promised . . .'*, J. Gould to C. Bonaparte, 30 Jan. 1850; Musée Nat. d'Hist. Nat., Paris

Exhibitor 247. 'Gould, J. XXVII (Fine Art Court), 247', Royal Commission 1851-2. Official, descriptive and illustrated catalogue, (4 vols.; London, Spicer Bros, 1851-2), vol.II, p.836

'Numerous attempts . . .', Monograph of the Trochilidae, or Family of Humming-birds (preface), 1861

'Mr. Hart commenced working . . .', B. Sharpe, *op. cit.*, p.xxi

Page 165 *'It has always been represented . . .'*, W. L. Baily ('William Loyd Baily, Sr') *Cassinia* (1919), No. 23:1-13

Page 166 W.L. Baily, Phila., to J. Gould 31 Aug. 1854; Zoo. Lib., Nat. Hist. Mus., London

Page 167 J. Gould to W.L. Baily, Phila., 19 Oct. 1854; *Cassinia, loc. cit.*, Lib. of Acad. Nat. Sci., Phila.

W.L. Baily to J. Gould, 31 March 1855; Zoo. Lib., Nat. Hist. Mus., London

Page 168   J. Gould to W.L. Baily, 8 Aug. 1855, *Cassinia, loc. cit.*, Lib. of Acad. Nat. Sci., Phila.

             W.L. Baily to J. Gould, 14 April 1856; Zoo. Lib., Nat. Hist. Mus., London

             J. Gould to W.L. Baily, 4 June 1856; *Cassinia, loc. cit.*, Lib. of Acad. Nat. Sci., Phila.

Page 169   G. Lawrence to W.L. Baily, 10 Nov. 1855, *Cassinia, op. cit.*, p.9; Acad. Nat. Sci., Phila.

             Lizzie Gould to Louisa Gould, 3 May 1856; Dr. G. Edelsten

             'The drawing room when first I . . .', Eliza Gould Muskett Moon. 'Some pages of reminiscences of her early life' (1885); private

Page 170   *'I shall never forget. . . .'*, W.H. Hudson, *Idle Days in Patagonia* (NY, Dutton, 1917), p.180

CHAPTER 13: Tresses of the Day-Star

Page 172   C. Dickens, *Household Words*, No. 65 (21 June, 1851), p.28-91

             J. Gould to members of the Zoo. Council, 5 Feb. 1851; Zoo. Soc. London

Page 174   Queen Victoria's Journal, manuscript, (1851); The Roy. Lib., Windsor Castle

Page 175   *'the cases were placed . . .'*, B. Sharpe, *op. cit.*, p.xx

             'tremulous as when . . . wilderness'; 'revolving cases', The *Times*, 3 Sept. 1851, (Emu 21:114-125)

Page 176   J. Gould to Council of the Zoo. Soc., 19 Nov. 1851; Zoo. Soc., London

             J. Gould to R.J. Shuttleworth, 9 April 1851; Zoo. Lib., Nat. Hist. Mus., London

CHAPTER 14: Sir, the House is Unhealthy

Page 179   C. Dickens, *Nicholas Nickleby*, (1839)

             'One of our greatest amusements . . .', Eliza Gould Muskett Moon, 'Some pages of reminiscences of her early life' (1885); private

Page 180 E. Prince to J. Gould, 20 March 1839; Mitchell Lib., Sydney

Page 180 E. Prince to J. Gould, 15 Jan. 1840; Mitchell Lib., Sydney

Lizzie's Gould's journal, manuscript; Dr. G. Edelsten

Page 182 J. Snow, Snow on cholera, being a reprint of two papers by John Snow, MD together with a biographical memoir by B W Richardson, MD and an introduction by Wade Hampton Frost, MD (NY, Commonwealth Fund, 1936), p.52-3

Page 184 J.H. Gould to J. Gould, 7 Nov. 1854, *Letters from an Assistant-Surgeon in the Honourable East India Company's Service, to his father*, (London, privately printed)

Page 185 J. Gould to C. Bonaparte, 8 Dec. 1855, Musée Nat. d'Hist. Nat., Paris

CHAPTER 15: In Quest of a Living Hummer

Page 187 J. Gould, New York, to his children, 16 May 1857; Spencer Lib. Univ. Kansas, MS P463A:1

Page 189 *'actions appeared unlike anything . . .'*, J. Gould, *Trochilidae* (Introduction), *The Humming Birds* (Ridgway 1861), p.272

'It was a bold man . . .', A. Newton, *The Humming Birds*, (Ridgway), p.268

'The almost total absence . . .', J. Gould, *Monograph of the Trochilildae*, vol.3, pl.131

Page 190 J. Gould, *Trochilidae* (Introduction), (Ridgway, 1861), p.277

C. Gould, Boston, to S. Gould, 11 June 1857; Dr G. Edelsten

Page 193 *'the little creatures . . .'*, Latham, (Ridgway), p.277

'The specimens I brought alive . . .', J. Gould, *Trochilidae, loc. cit.*

Page 194 *'In passing through this world . . .'*, J. Gould, *A Monograph of the Trochilidae* (Preface)

CHAPTER 16: **Wolf, Ruskin, and the Birds of Great Britain**

Page 195 'I am living still nearer . . .', J. Gould to Dr H.l Schlegal. 19 Oct. 1860; Rijks. Nat. Hist., Leiden

J. Gould to A. Newton, 22 March 1861; Balfour & Newton Lib., Cambridge Univ.

Page 196 'I am sorry for the mistake . . .', A. Newton to J. Gould, 26 March 1861; Balfour & Newton Lib., Cambridge Univ.

'I am now busily engaged . . .', J. Gould to Sir R. Barry, 22 May 1863; State Lib., Melbourne, Vict.

A.H. Palmer, *Life of Joseph Wolf*, (London & New York, Longmans, Green & Co., 1895) pp.342-52

Page 202 J. Ruskin, E.T. Cook & A. Wedderburn (eds.), *Works of Ruskin*, vol.XXIV, 1903-12, p.44-5

Page 203 'Love's Meinie', Lecture III, J. Ruskin, E.T. Cook & A. Wedderburn (eds.), *op. cit.*, vol.XXV (1903-12)

Page 204 J. Gould to F. McCoy, 20 Dec. 1865; Nat. Mus. Vict., Australia

Page 205 J. Gould to anon., 14 Dec. 1868; Ellis Collection, Spencer Lib., Univ. Kansas, MS P463A:5

Page 207 C. Gould to L. Gould, 23 Jan. 1882; Dr G. Edelsten

CHAPTER 17: **The Curtain Falls**

Page 208 'I have almost all the colourers. . . .', J. Gould to G.N. Lawrence, New York, 22 June 1871; Amer. Mus. of Nat. Hist., letter no. 25a

'I am sorry to hear . . .', F. Gould to J. Gould, 19 Nov. 1869; Duke of Westminster

Page 209 B. Sharp, *op. cit.*, p.xxiv

Eyewitness account of F. Gould's death, by W.F. Creeny, MA, 20 March 1873; Dr G. Edelsten

Page 210 S. Gould to L. Muskett, 24 March 1873; Duke of Westminster, Eaton, Chester

'he had a really tender . . .', B. Sharpe, *op. cit.*, p.xxiii

'We, as a rule . . . profer more', C. Gould to his sisters, 23 Dec. 1884; Dr G. Edelsten

Page 211 *'He hated the sight of a stranger. . . .'*, J.G. Millais, *The Life and Letters of Sir John Everett Millais*, (Methuen & Co., 1899), p. 169

A.A. Prince to G.N. Lawrence, 3 Sept. 1873; Amer. Mus. of Nat. Hist., letter no. 6

B. Sharpe, *op. cit.*, p.xxi

Page 212 J. Gould to G.N. Lawrence, 12 April 1876; Amer. Mus. of Nat. Nist., letter no. 9

J. Gould to F. McCoy, 15 April 1876; Nat. Mus. of Vict., Australia

J. Ruskin to Rev. J.P. Faunthorpe, 9 Dec. 1881, Cook & Wedderburn (eds.), *op. cit.*, vol.XXXVII, Letters of Ruskin, p.381

Page 213 M. Roberts, *W.H. Hudson, a portrait*, (London, Eveleigh, Nash & Grayson, 1924), pp.29-30

Page 214 J.G. Millais, *op. cit.*, pp.169-73

'there are only three things. . . .', J. Ruskin, Cook & Wedderburn (eds.), *op. cit.*, vol.XIV, appendix X, p.496

Page 215 *'I have never seen . . .'*, J.G. Millais, *op. cit.*, p.173

'I have made a great mistake . . .', J. Ruskin, Cook & Wedderburn (eds.), *op. cit.*, vol.XXXIV, p.670

Page 216 *'I once asked him . . .'*, B. Sharpe, *op. cit.*, p.xxiv

J. Gould's will, manuscript, 1878, Somerset House, London

'I am sorry to say . . .', J. Gould to G.N. Lawrence, 1 Jan. 1878; Amer. Mus. of Nat. Hist., letter no. 13

J. Gould to J. Noble, 17 Oct. 1878; The Nat. Soc., Griggsville, IL

Page 217 *'By hook or crook . . .'*, J. Gould to J. Noble, 22 Oct. 1878; The Nat. Soc., Griggsville, IL

J. Gould to R.B. Sharpe, 7 June 1879; Blacker-Wood Lib. of Biology, McGill Univ., Montreal

Page 218 D.G. Elliot, Paris, to R. Ridgway, 29 March 1878; Ridgway Collection, McGill Univ., Montreal

D.G. Elliott, New Brighton, to R. Ridgway, 9 Jan. 1879; Ridgway Collection, McGill Univ., Montreal

Page 219 J. Gould to G.F. Westerman, Amsterdam, 9 Dec. 1879; Gemeentearchief, Amsterdam

J. Gould to R.B. Sharpe, 8 Sept. 1880; Blacker-Wood Lib. of Biology, McGill Univ., Montreal

as an old friend . . .', S. Gould to R.B. Sharpe, 4 Feb. 1881; Blacker-Wood Lib. of Biology, McGill Univ., Montreal

T. Salvadori, Turin, to R.B. Sharpe, 24 July 1881, Blacker-Wood Lib. of Biology, McGill Univ., Montreal

J. Gould to G.F. Westerman, Amsterdam, 9 Dec. 1879; Gemeentearchief, Amsterdam

of several. . . .', H. Sotheran & Co., London, to Sec. of the Aust. Mus., 17 Jan. 1884; Nat. Mus. of Aust., Sydney

marvellous works. . . .', A. Newton, *A Dictionary of Birds,* (London, A & C Black, 1896)

Page 220 *'Memoir of the Late John Gould',* Zoologist, third series, vol.v (1881), p.115

he had left. . . . ability', Duke of Westminster to Miss Gould, 5 Feb. 1881; Dr. G. Edelsten

J. Gould's will, manuscript, 1878, Somerset House, London

Here lies John Gould . . .', B. Sharpe, *op. cit.,* p.xxiv

Select Bibliography

Allen, David Elliston, *The Naturalist in Britain*, Penguin Books, 1976

Barber, Lynn, *The Heyday of Natural History*, Garden City, NY, Doubleday & Co., 1980

Blunt, Wilfrid, *The Ark in the Park*, Hamish Hamilton, 1976

Bryant, C.E., ed., 'Gould Commemorative Issue', *Emu*, vol. 38 (1 Oct. 1938), pt.2, pp.89-244, Melbourne, Royal Australasian Ornithologists Union

Chisholm, A.H., *Strange New World. The Adventures of John Gilbert and Ludwig Leichhardt*, Sydney, Angus & Robertson, 1941

  The Story of Elizabeth Gould, Melbourne, Hawthorne Press, 1944

Colp, Ralph Jr., 'Charles Darwin and the Galapagos', *NY State J. of Med*. 77 (1977):262-7

Corning, Howard, *Letters of John James Audubon*, 1826-40, 2 vols, Boston, 1930

Darwin, Charles, *Voyage of the Beagle*, 1839

  The Origin of Species, 1859

Dawes, Sean C., 'Towards identifying the first travellers overland from Adelaide to the North West Bend of the River Murray', *Proc. Roy. Geog. Soc. Australasia, S. Austral. Branch*, 81 (1980-1):79-85

Farber, Paul Lawrence, *Emergence of Ornithology as a Scientific*

Discipline: 1760-1850, Dordrecht, Holland, D. Reidel, 1982

Ford, Alice, *John James Audubon*, Oklahoma Univ. Press, 1964

Hyman, Susan, *Edward Lear's Birds*, London, Weidenfeld and Nicholson, 1980

Jackson, Christine, *Bird Illustrators*, London, Witherby, 1975

Jardine, William and Selby, Prideaux John, *Illustrations of Ornithology*, 1825-43, 4 vols, Edinburgh, Lizars

Knight, David, *Zoological Illustration*, Folkestone, Kent, William Dawson, 1977

Lack, David, *Darwin's Finches*, Cambridge Univ. Press, 1947

Longmate, Norman, *King Cholera – the Biography of a Disease*, London, Hamish Hamilton, 1966

Martin, W.C.L., *A general history of Humming-birds or the Trochilidae: with especial reference to the collection of J. Gould, F.R.S. etc now exhibiting in the gardens of the Zoological Society of London*, London, H.G. Bohn, 1852

McEvey, Allan, *John Gould's Contribution to British Art*, Sydney Univ. Press, 1973

Merrill, Lynn C., *The Romance of Victorian Natural History*, Oxford Univ. Press, 1989

Mitchell, David William, 'Gould's birds', *Westminster Review*, 35 (1841):271-303

Mitchell, P. Chalmers, *Centenary History of the Zoological Society of London*, London, Zoological Society, 1929

Noakes, Vivien, *Edward Lear – the Life of a Wanderer*, London, Collins, 1968

Selected Letters of Edward Lear, Oxford Univ. Press, 1988

Owen, R., 'The Giraffe', *Zoological Magazine* 1 (1833):1-14

Palmer, A.H., *The Life of Joseph Wolf, Animal Painter*, London and New York, Longmans, Green & Company, 1895

Parker, Shane A., 'Remarks on some results of John Gould's visit to South Australia in 1839', *South Australian Ornithologist* 29 (1984):109-12

Reade, Brian, *Edward Lear's Parrots*, London, Duckworth, 1949

Reeve, Lovell (ed.), *Portraits of Men of Eminence in Literature, Science, and Art, with Biographical Memoirs*, vol.2, London, Reeve & Co., (1863-7)

Ridgway, Robert, 'The Humming Birds', Report of National Museum, pp.253-383, 1890

Sauer, Gordon C., *John Gould, the Bird Man. A Chronology and Bibliography*, Landsdowne Editions, Melbourne, 1982

'John Gould in America', *Contributions to the History of N. American Natural History*. London, Soc. for the Bibliog. of Nat. Hist., 1983

Scherren, Henry, *The Zoological Society of London*, London, Cassell, 1905

Sharpe, R. Bowdler, *An Analytical Index to the Works of the late John Gould, FRS*, London, Henry Sotheran, 1893

Spinage, C.A., *The Book of the Giraffe*, London, Collins, 1968

Stresemann, Erwin, *Ornithology from Aristotle to the Present*, Cambridge, Harvard Univ. Press, 1975

Sulloway, Frank J. 'Darwin's Conversion: The Beagle Voyage & Its Aftermath', *J. Hist. Biology*, vol. 15 (1982), no.3, pp.325-96

Sulloway, Frank J., 'Darwin's Finches', *Bull. Br. Mus. Nat. Hist.* (Zool.) 43(2) (1982):49-94

Swainson, William, *Taxidermy, with the Biographies of Zoologists and Notices of their Works*, London, Longman, Orme, Brown, Green & Longman's, 1840

Swainson, William, *Preliminary Discourse on Natural History*, London, Longman, Orme, Brown, Green & Longman's, 1834

Tate, Peter, *A Century of Bird Books*, London, Witherby, 1979

Wagstaffe, R. and Rutherford, G., *Letters from Knowsley Hall, Lancashire*, North Western Naturalist, 1954-5

Whittell, H.M., 'A review of the work of John Gilbert in Western Australia', *Emu* 41 (1941): pp.112-29, 216-42, 289, 305; *Emu* 51 (1951):17-29

Whittell, H.M., *The Literature of Australian Birds*, Perth, Paterson Brokensha, 1954

The Illustrated Folio Works of John Gould

1830-1   *A Century of Birds from the Himalaya Mountains* 1 volume, 80 plates; London

1832-7   *The Birds of Europe* 22 parts, 5 volumes, 448 plates; London

1833-5   *A Monograph of the Ramphastidae, or Family of Toucans* 3 parts, 1 volume, 34 plates; London

1836-8   *A Monograph of the Trogonidae, or Family of Trogons* 3 parts, 1 volume, 36 plates; London

1837-8   *Icones Avium, or Figures and Descriptions of New and Interesting Species of Birds from Various Parts of the World* 2 parts, 18 plates; London

1837-8   *The Birds of Australia and the Adjacent Islands* 2 parts, 20 plates; London (later suppressed)

1840-8   *The Birds of Australia* 36 parts, 7 volumes, 600 plates; London

1841-2   *A Monograph of the Macropodidae, or Family of Kangaroos,* 2 parts, 30 plates; London

1844-50   *A Monograph of the Odontophorinae, or Partridges of America,* 3 parts, 1 volume, 32 plates; London

1849-61   *The Mammals of Australia,* 13 parts, 3 volumes, 182 plates; London

1850-83   *The Birds of Asia* 35 parts, 7 volumes, 530 plates; London (completed by R. Bowdler Sharpe)

1851-69   *Supplement to the Birds of Australia,* 5 parts, 1 volume,

81 plates; London

1852-4   *A Monograph of the Ramphastidae, or Family of Toucans*, 2nd edition, 3 parts, 1 volume, 52 plates; London

1855   *Supplement to the first edition of 'A Monograph of the Ramphastidae, or Family of Toucans'*, 2 parts, 20 plates; London

1858-75   *A Monograph of the Trogonidae, or Family of Trogons*, 2nd edition, 4 parts, 1 volume, 47 plates; London

1862-73   *The Birds of Great Britain* 25 parts, 5 volumes, 367 plates; London

1875-88   *The Birds of New Guinea, and the Adjacent Papuan Islands* 25 parts, 5 volumes, 320 plates; London (completed by R. Bowdler Sharpe)

1880-7   *Supplement to the 'Monograph of the Trochilidae, or Family of Humming-birds'* 5 parts, 1 volume, 58 plates; London (completed by R. Bowdler Sharpe)

1880   *A Monograph of the Pittidae, or Ant-Thrushes of the Old World*, 1 part, 13 plates; London (incomplete)

Additional Works

1837-8   *A Synopsis of the Birds of Australia and the Adjacent Islands* imperial octavo, 4 parts, 73 plates; London

1848   *An Introduction to the Birds of Australia* octavo; London (printed for private circulation)

1861   *An Introduction to the Trochilidae, or Family of Humming-birds* octavo; London

1863   *An Introduction to the Mammals of Australia* octavo; London (printed for private circulation)

1865   *Handbook to the Birds of Australia* 2 volumes, octavo; London

1873   *An Introduction to the Birds of Great Britain* octavo; London

Index